ADA & BUILDING TRANSPORTATION

A Handbook on Accessibility Regulations for Elevators, Wheelchair Lifts and Escalators

Edward A. Donoghue, CPCA, President
Edward A. Donoghue Associates, Inc.
Code and Safety Consultants
Salem, New York

Copyright ©2000 by ELEVATOR WORLD. All rights reserved. Printed in the U.S. No part of this publication may be reproduced without prior written permission of the publisher. *ADA and Building Transportation* is available from Elevator World, Inc., Educational Division, P.O. Box 6507, Mobile, AL 36660.

IMPORTANT NOTICE – Please Read Carefully

This second edition of *ADA and Building Transportation* (formally *ADA & Vertical Transportation*) is published by Elevator World Inc., as a service to the building transportation industry. References to cited codes, standards and regulations are intended to be read in conjunction with the applicable referenced documents. The cited documents and the interpretation of each may be subject to future change as a result of legislation or litigation or in the interest of clarification.

This information is presented without warranty, and neither The Publisher, Elevator World Inc. nor the author, Edward A. Donoghue, president of Edward A. Donoghue Associates Inc. (hereafter EADAI) nor their officers, or employees assume any liabilities or responsibilities for the accuracy of information, comments, opinions, representation or data contained herein. No liability is assumed by Elevator World Inc. and EADAI for decisions or commitments made on the basis of the information provided.

The American with Disabilities Act (ADA) provides for injunctive relief, and in some cases monetary fines to be levied, against a noncomplying facility. A person held to have been discriminated against under the Act may in some instances be entitled to monetary compensation from the entity in violation. Therefore, it is recommended that counsel be consulted before decisions are made concerning proposed actions to be taken in efforts to comply with ADA.

About the Author

Edward A. Donoghue, CPCA is president of Edward A. Donoghue Associates Inc., Code and Safety Consultants, Salem, New York. Mr. Donoghue is a member of the American National Standards Institute Committee A117 on Architectural Features and Site Design of Public Buildings and Residential Structures for Persons with Handicaps. He is Chairman of the American Society of Mechanical Engineers (ASME) Qualification of Elevator Inspectors Committee and in 1986-1987 chaired the Safety Division of ASME. He was the 1992 recipient of the ASME Safety Codes and Standards Medal.

Mr. Donoghue is also a member of the ASME A17 Safety Code for Elevators and Escalators Main, Hoistway, Emergency Operations, Construction Elevator, Existing Elevator, Dumbwaiter, Inspectors' Manual, International Standards, Sidewalk Elevator, Limited Use/Limited Application Elevator and Evacuation Guide Committees and chairs the Editorial and Code Coordination Committees. He is also an associate member of the Canadian Standards Association (CSA) B44 Elevator Safety Code Committee. Mr. Donoghue was a licensed master electrician and building contractor. He is the author of the "Handbook A17.1 Safety Code for Elevators and Escalators."

He has been certified as a Professional Code Administrator by the National Academy of Code Administration and as an Elevator Inspector by NAESA, BOCA, and SBCCI. Mr. Donoghue is a member of American Society of Mechanical Engineers (ASME), International Association of Elevator Engineers (IAEE), American Society for Testing and Materials (ASTM), Building Officials and Code Administrators International (BOCA), Southern Building Code Congress International (SBCCI), International Conference of Building Officials (ICBO), National Association of Elevator Contractors (NAEC), National Association of Elevator Safety Authorities (NAESA), National Fire Protection Association (NFPA), and National Conference of States on Building Codes and Standards (NCSBCS).

Acknowledgements

I am indebted to George W. Gibson, formerly director of Codes and Product Safety, Otis Elevator Company, and now the principal of George W. Gibson & Associates, Inc., who encouraged me to write this book and for reviewing the manuscript for the first edition. My thanks are also extended to Brian Black, director of Building Codes and Standards, Eastern Paralyzed Veterans Association who was the recipient of numerous telephone calls while researching the material for the Handbook and for his time reviewing the manuscript for both the first and second editions. Thanks are also extended to Zack McCain, Jr., P.E., President McCain Engineering Associates Inc., who took time from his busy schedule to review the manuscript for the first edition. I also wish to thank Ralph Droste, manager, Codes and Electrical Product Safety, Otis Elevator Company, who contributed the majority of Chapter 5 and George Kappenhagen, code consultant, Schindler Elevator Corporation, for the drawings illustrating wheelchair reach requirements applicable to elevator car controls and information on destination oriented elevators.

I owe a great deal to National Elevator Industry, Inc. for giving me the opportunity to represent their interest at the accessibility codes and standards writing organizations. My good friends Elmer Sumka, now deceased, and Robert Young over the years freely shared their knowledge on the reasons and research that led to today's accessibility regulations.

Finally, I must thank my wife and partner, Janet M. Donoghue, who labored typing, proofreading and retyping the manuscript for this Handbook.

Foreword

A question often asked is what accessibility standard do I follow; ADAAG, FHAA, MGRAD, UFAS, ANSI A117 or local requirements? Many are of the mistaken impression they must only comply with ADAAG. While ADAAG has taken center stage it does not apply universally.

The model building code organizations have been working to regain control of accessibility regulation ever since ADA was enacted. The three model building codes in the fall of 1992 enacted provisions for their 1993 Codes, which should accomplish that goal. All have adopted accessibility scoping provisions and the CABO/ANSI A117.1-1992 Standard. The 1999 edition of NBC and 1999 Supplement to the SBC have adopted ICC/ANSI A117.1-1998. In reality, the model building codes are probably more stringent than ADAAG, in many areas.

Codes, standards and regulations must be written for enforcement purposes. As such, the text is concise, without examples and explanations. ADA and Building Transportation contains rational for the requirements; explanations, examples and excerpts from the codes, standards and regulations that are referenced. The information was compiled from committee minutes, correspondence and interpretations, as well as conversations with past and present personnel involved with the development, implementation, enforcement and interpretation of accessibility regulations.

The original intent for many of the requirements is obscure in the records. The author therefore has tried to convey, through text and illustrations, the results of the requirements as applied to equipment installed today. It should not be construed, however, that the examples and illustrations are the only means for complying with the requirements. With information of this type it is hoped that the reader of ADA and Building Transportation will have a better comprehension of, and appreciation for, the various accessibility requirements.

The material in ADA and Building Transportation is the opinion of the author, and does not necessarily represent an official opinion of the organization responsible for the development and/or enforcement of the regulation, code or standard in question. When an official interpretation is needed the reader should address their question, in writing, to the authority having jurisdiction.

Since 1993 when the first edition of ADA and Building Transportation was published the Access Board and A117 Committee have had numerous meetings to further refine, clarify and harmonize their respective documents. The Access Board appointed an ADAAG Review Advisory Committee to review the differences between CABO/ANSI A117.1 and ADAAG. They issued a final report with recommendations for a new ADAAG in September 1996. This report is the basis for many of the revisions that appear in ICC/ANSI A117.1-1998. The Access

Board is also planning to revise ADAAG. The stated objective of both organizations is harmonization of requirements.

In the spring of 1999, Elevator World's stock of the first edition was quickly being exhausted. Do I wait for the new ADAAG or update ADA and Building Transportation to reflect current accessibility regulations? The Access Board originally planned to release a draft revision to ADAAG in the spring of 1997. At the time this is being written the draft is scheduled to be published in the Federal Register in the late fall 1999. I concluded that a revised ADAAG would not be approved and published for at least one year and more likely two years. Then it must be incorporated into the ADA regulations promulgated by DOJ. In the meantime the readers of ADA and Building Transportation will be faced with complying with ICC/ANSI A117.1-1998. The decision was made that a second edition was needed at this time. If and when a revised ADAAG is enacted into the ADA regulations, a third edition of ADA and Building Transportation will be written.

AS WE GO TO PRESS

The Access Board on November 16, 1999 published in the Federal Register the first draft of ADA/ABA Accessibility Guidelines. This document is intended to eventually replace ADAAG and UFAS. See Appendix C of this Handbook for excerpts from ADA/ABA Accessibility Guidelines, which should be of interest to the building transportation industry.

ABBREVIATIONS

The following is a list of the abbreviations for the terms that appear in this Handbook:

ABBREVIATION   **UNIT**

ADA	Americans with Disabilities Act
ADAAG	Americans with Disabilities Act Accessibility Guidelines
AEMA	Accessibility Equipment Manufacturers Association
ANSI	American National Standards Institute
ASME	American Society of Mechanical Engineers
ATBCB	Architectural and Transportation Barriers Compliance Board
BCMC	Board for the Coordination of the Model Codes
BOCA	Building Officials and Code Administrators International
CABO	Council of American Building Officials
CSA	Canadian Standards Association
deg.	degree (angle)
DOJ	Department of Justice
DOT	Department of Transportation
FHAA	Fair Housing Amendments Act
ft	feet
HUD	Department of Housing and Urban Development
Hz	hertz
IBC	International Building Code
ICBO	International Conference of Building Officials
ICC	International Code Council
in.	inch
lb	pound
LU/LA	limited-use/limited-application
mm	millimeter
NBC	National Building Code
NEII	National Elevator Industry, Inc.
NFPA	National Fire Protection Association
NIST	National Institute of Standards and Technology
SBC	Standard Building Code
SBCCI	Southern Building Code Congress International
sec	second (time)
UBC	Uniform Building Code
UFAS	Uniform Federal Accessibility Standards

Table of Contents

Important Notice ii
About the Author iii
Acknowledgments iv
Foreword v
Abbreviations vii
Table of Contents viii

Chapter 1 — ACCESSIBILITY REQUIREMENTS - PAST, PRESENT AND FUTURE
    1.1   ANSI A117.1 1
    1.2   NEII Handicapped Standard 3
    1.3   Uniform Federal Accessibility Standard 4
    1.4   Fair Housing Amendments Act 5
    1.5   Americans with Disabilities Act 6
    1.6   ADAAG - The Future 10
    1.7   Limited Use/Limited Application Elevators 12
    1.8   Evacuation of Disabled in a Fire 15

Chapter 2 — OVERVIEW OF ADA TITLE III REGULATIONS
    2.1   ATBCB's Accessibility Guidelines 19
    2.2   General 19
    2.3   Existing Facilities 20
    2.4   New Construction 22
    2.5   Effective Dates 23
    2.6   Elevators Required 24
    2.7   Elevators Not Required 24
    2.8   Enforcement and Legal Remedies 25
    2.9   State and Local Accessibility Requirements 28

Chapter 3 — WHAT ACCESSIBILITY REGULATIONS APPLY
    3.1   ADA 33
    3.2   Fair Housing Amendments Act 34
    3.3   Uniform Federal Accessibility Standards 35
    3.4   Model Building Codes 35
    3.5   Accessibility Regulations by Building Occupancy 38

Chapter 4 — REVIEW OF REGULATIONS
    4.1   Analysis of Elevator Regulations By Section 43
    4.1.1   General 43
    4.1.2   Automatic Operation 45
    4.1.3   Call Buttons 47
    4.1.4   Hall Signals 49
    4.1.5   Tactile Characters on Hoistway Entrances 51
    4.1.6   Doors 52

4.1.7	Door and Signal Timing for Hall Calls	54
4.1.8	Door Delay for Car Calls	56
4.1.9	Inside Dimension of Elevator Cars	57
4.1.10	Floor Surfaces	62
4.1.11	Illumination Levels	63
4.1.12	Car Controls	64
4.1.13	Car Position Indicators	70
4.1.14	Emergency Communications	72
4.1.15	Transportation Facilities	74
4.2	Analysis of Wheelchair Regulations by Section	75
4.2.1	Location	75
4.2.2	Requirements	77
4.2.3	Entrance and Operation	81
4.3	Analysis of Escalator Regulations	82
4.4	Destination Oriented Elevators	84
4.4.1	Call Buttons	84
4.4.2	Hall Signals	85
4.4.3	Tactile Characters on Hoistway Entrances	86
4.4.4	Car Controls	86
4.4.5	Car Displays	86
4.5	Maintenance	86
4.6	Private Residence	88
4.7	Maintenance	88

Chapter 5 — ADAAG QUESTIONS AND ANSWERS 93

Chapter 6 — VERTICAL TRANSPORTATION ACCESSIBILITY STANDARDS COMPARISON CHART 107

Chapter 7 — ADAAG - ELEVATOR CHECKLIST 135

        APPENDIX A - REFERENCE MATERIAL 145

        APPENDIX B - SIGNAGE STANDARDS 148

        APPENDIX C – DRAFT ADA/ABA ACCESSIBILITY GUIDELINES BUILDING TRANSPORTATION REQUIREMENTS 156

ACCESSIBILITY REQUIREMENTS
Past Present & Future

1.

1. ACCESSIBILITY REQUIREMENTS – PAST, PRESENT AND FUTURE

1.1 ANSI A117.1

In May of 1959, the American Standards Association, now known as the American National Standards Institute (ANSI), at the request of The President's Committee on Employment of the Physically Handicapped, sponsored a conference with those groups interested in accessibility. The conference participants concluded that an accessibility standard was needed. The President's Committee on Employment of the Physically Handicapped and the National Society for Crippled Children and Adults, now known as the National Easter Seal Society, were designated cosponsors and later agreed to assume the Committee Secretariat. The Committee developed the first ANSI A117.1 Standard, which was approved October 31, 1961. Some 48 organizations were represented on the original committee. National Elevator Manufacturing Association, now known as National Elevator Industry, Inc. (NEII), being one. The elevator requirements, like the rest of the standard, were minimal. Included were requirements that elevators are needed in multistory buildings and they should serve all levels normally used by the general public. The requirement concluded "that elevators shall allow for traffic by wheelchairs" and referred to general principles and considerations for wheelchair use. Those included wheelchair dimensional specifications, turning space and reach requirements. In hindsight, this standard was far from adequate. In spite of this, it was reaffirmed in 1971. That action resulted in an outcry to produce a standard that realistically addressed the needs of the disabled.

In 1974, the United States Department of Housing and Urban Development (HUD) joined the Secretariat and sponsored a research and development project. The research grant was awarded to Edward Steinfeld, a professor of architecture at Syracuse University. Edward Steinfeld also was appointed Secretary of the ANSI A117 Committee. The research program began in 1974 and NEII was contacted to provide input for the elevator requirements. Robert Young (Otis Elevator Company) and Elmer Sumka (Westinghouse Elevator Company) were appointed to represent NEII. Countless meetings with Syracuse University, A117, HUD and other interested parties followed. Meetings were also held throughout the U.S. with the disabled community and their representative organizations. As a result, the accessibility requirements for elevators evolved, but not without discussion, disagreement and compromise. The bases for many of the compromises are detailed in Chapter 4. By early 1976, the elevator provision for ANSI A117.1 had generally been agreed to. The other provisions (e.g. ramps, doors, toilet facilities, etc.) took four more years and it was not until 1980 that the revised ANSI A117.1 was approved and published. Shortly after the publication of the 1980 edition of ANSI A117.1, the National Easter Seal Society was recognized by ANSI as the Administrative Secretariat of the Committee. To encourage the member's participation, the A117 Secretariat established five task forces to review and recommend revision to the 1980 edition. Elmer Sumka and George Gibson represented NEII on the ANSI A117 Committee and task force responsible for the elevator provision. The recommendation from the task forces formed the basis for the next edition of ANSI A117.1 that was issued in 1986.

During the late 1980s, the National Easter Seal Society decided that they did not wish to continue as the Administrative Secretariat. HUD and The President's Committee on Employment of the Physically Handicapped also did not wish to continue. A search for a new Secretariat commenced with the Council of American Building Officials (CABO) assuming sole responsibility for the ANSI A117 Committee. CABO initiated a revision process in July 1989. From the outset the primary goal of the Committee was to produce a standard that would be compatible with model building codes. Attainment of that goal would provide building code enforcement authorities with enforceable criteria. Enforceable criteria would provide superior accessibility to buildings and facilities for persons with disabilities. Throughout the revision process, the 44 member Committee endeavored to eliminate application criteria such as; scoping requirements that tell where, when, and to what extent the accessibility requirement will apply to the built environment and unenforceable, informational statements that explain the reason for the criteria.

To this end, the Committee appointed an editorial subcommittee with assignments to:
- make the format consistent throughout;
- identify and remove unenforceable informational statements;
- delete application criteria (scoping requirements that are contained in building codes); and
- write criteria contained in figures into the text of the standard.

Informational statements removed from the text were, as appropriate, transferred to the appendix that is not part of the Standard. Because the figures are no longer used to establish criteria, they were moved to an appendix also. This transfer of the figures to an appendix serves to emphasize that they are not part of the Standard and cannot, therefore, contain criteria different from that contained in the Standard. Rather, they are being used in the usual manner of illustrating application of the criteria.

The Author and John Herwig (General Elevator Company) represented NEII at the CABO/ANSI A117 Committee meetings. The 1992 edition of the CABO/ANSI A117.1 Standard reflects 3½ years of work to produce an enforceable document that will bring about greater accessibility for people with disabilities.

The CABO Board for the Coordination of the Model Codes (BCMC), concurrent with the revision to the CABO/ANSI A117-1992 standard, developed model application criteria (scoping requirements) which have been incorporated into model building codes. This coordinated effort gave the authority having jurisdiction a compatible set of application and technical criteria.

In September 1994, the CABO/ANSI A117 Committee met in Baltimore, Maryland and decided it was time to start another revision cycle. Public input was solicited and over five hundred proposals were received. The A117 Committee also reviewed the ADAAG Review Federal Advisory Committee final report, with the intended objective of publishing a standard that was harmonized with a planned revision to ADAAG. Unfortunately, a revised ADAAG was delayed and that objective was not obtained. The CABO/ANSI A117 Committee also put considerable effort into an editorial reformatting of the document. Three years later, the revision was approved by ANSI on February 13, 1998 and published in October 1998. After approval by ANSI and prior to publication, the Secretariat CABO was assimilated into the International Code Council (ICC) the document thus being identified as ICC/ANSI A117.1-1998. ICC/ANSI A117.1-1998 was developed to work in harmony with Federal accessibility laws, including the Fair Housing Accessibility Guidelines and the Americans with Disabilities Act Accessibility Guidelines (ADAAG). Highlights include expanded sections on elevators, signs, alarms and Automatic Teller Machines, a revised unobstructed high side reach requirement, and new criteria for detectable warnings for platform edges. In addition, the standard has been completely reformatted to simplify its use, and illustrative figures are provided throughout the document. Representing NEII on the A117 Committee for this cycle was your author with Ralph Droste of Otis Elevator Company and George Kappenhagen of Schindler Elevator Corporation being my alternates. Also joining the Committee was AEMA with Patrick Bass of National Wheel-O-Vator Company acting as their representative.

1.2 NEII HANDICAPPED STANDARD

In the early 1970s, accessibility was an issue that most jurisdictional authorities were just starting to address. The only national standard available was the 1961 edition of A117.1, which was generally recognized as inadequate. Work had begun in 1974 on a new edition of A117 but local authorities could not wait for that project to be completed. The elevator industry by 1975 was facing a myriad of different accessibility regulations often with conflicting requirements. The research for the 1980 edition of A117.1 on elevators was completed by early 1976. Agreement on the elevator requirements for the next edition was generally obtained. NEII recognized that the Committee would not reach consensus on the other areas covered in A117.1 for a number of years. The elevator industry could not afford the luxury of waiting. NEII set out to publish an accessibility standard that met the needs of the elevator industry.

The NEII standard was the first modern set of requirements for making elevators usable by and accessible to the disabled. The "NEII Minimum Passenger Elevator Requirements for the Handicapped" was first published in 1976. Over the years, four editions were published. The latest edition was published in July 1985. NEII's goal was to provide accessible elevators utilizing standard equipment wherever possible to provide for safe, convenient and acceptable service for all users at a minimum increase in cost. The standard was well received and formed the basis for many state and local regulations.

In the early 1980s, NEII became aware that its standard was also being applied to existing installations. NEII recognized that many elevators were designed and installed in accordance with industry standards and code requirements, which predated current accessibility standards. It further recognized that this equipment might never be able to comply with current accessibility requirements. They also were aware that current requirements for new elevators were never intended to apply retroactively. However, the NEII Central Code Committee saw a need for making existing elevators accessible to the most practical extent possible. The Committee set out to develop requirements that would provide reasonable access, but not necessarily provide the same convenience as would be found on new installations. For example, it would neither be practical nor possible to prescribe that existing car sizes comply with those applicable to new elevators. Since, in many cases, hoistway walls would have to be moved to accommodate the newer sizes. This effort resulted in the 1985 edition of the "NEII Minimum Passenger Elevator Requirements for the Handicapped" including provisions to recognize both new and existing accessible elevators. Similar recognition of making existing elevators accessible was incorporated in CABO/ANSI A117.1-1992 and ICC/ANSI A117.1-1998. In the summer of 1991, NEII withdrew their standard as it was not in compliance with the recently published ADAAG.

1.3 UNIFORM FEDERAL ACCESSIBILITY STANDARD

The Architectural Barriers Act of 1968 and amendments have stipulated that all buildings designed, constructed, altered and (later) leased with federal financial assistance must be accessible according to a standard issued by federal agencies. The first standard applied was the early ANSI or equivalent. Today, the Uniform Federal Accessibility Standards (UFAS) is the applicable federal standard.

In Section 502 of the Rehabilitation Act of 1973, the Architectural and Transportation Barriers Compliance Board (ATBCB or Access Board) was established to ensure compliance with the standards issued under the 1968 Architectural Barriers Act. The Access Board is comprised of a unique mix of government agency representatives and private citizens, some with disabilities. The Access Board and technical staff members investigate and examine alternative approaches to the architectural, transportation, communication and attitudinal barriers confronting people with disabilities. The Access Board seeks voluntary compliance with UFAS from all federal agencies. If voluntary compliance is not forthcoming, the Board follows an administrative process and a set of enforcement rules to settle a dispute.

Authority was also given to the ATBCB to develop minimum guidelines for the federal agencies to use in developing standards. The ANSI A117.1-1980 technical provisions and structure were modified slightly and made part of the "Minimum Guidelines and Requirements for Accessible Design" (MGRAD). The MGRAD was mandated by Congress and developed by the Access Board for use by federal agencies in developing an improved and consistent federal standard. In addition to ANSI A117.1-1980 technical provisions, MGRAD added broad scoping requirements based on building types and occupancies, stipulating how many accessible features are required and where. MGRAD was thus the basis for UFAS, which was developed by the four standards setting organizations: General Services Administration (GSA); Department of Defense (DOD); United States Postal Service (USPS) and Department of Housing and Urban Development (HUD). The ATBCB was and still is a member of the ANSI Committee, and its representation provides an avenue for sharing research results and developing consistency between the federal government and private sector standards.

The Rehabilitation Act of 1973, Section 504 stipulated that any program or activity receiving any form of federal financial assistance had to be accessible to everyone. Section 504 and further amendments of the Rehab Act of 1973 have resulted in building modifications for accessibility because recipients of federal funds are required to make their programs accessible for people with disabilities by modifying existing facilities to make them accessible, by moving the program to an accessible facility, or by making other accommodations.

Early editions of ANSI A117.1 or equivalent standard, and its applicable provisions for access, were initially "adopted" by many agencies. Now, almost all federal agencies adhere to UFAS or ADAAG as the design standard to be followed for new construction, alterations or physical modification that may be needed to achieve Section 504 compliance. UFAS applies to federally owned and leased buildings. State and local governments that accept federal money have to comply with UFAS or ADAAG. If ADAAG is utilized, the elevator exemption does not apply. An objective of the Access Board is to sunset UFAS with the next edition of ADAAG being applicable to the federal, public and private sectors.

Public, private and religious institutions that accept federal money, even though the money is not used for construction, are also obligated to comply with UFAS. Some of these institutions may additionally be required to comply with ADAAG. (See Chapter 1, Section 1.5.)

1.4 FAIR HOUSING AMENDMENTS ACT

Title VIII of the Civil Rights Act of 1968 has prohibited discrimination in the sale, rental and financing of dwellings based on race, color, religion, sex or national origin. The Fair Housing Amendments Act, (FHAA) enacted September 13, 1988, expanded the coverage of Title VIII to prohibit discriminatory housing practices against persons with disabilities and families.

The regulations implementing FHAA prohibits discriminatory housing practices when based on race, color, religion, sex, disability, familial status or national origin. These prohibitions against discrimination apply to most dwellings, including condominiums and cooperatives. However, the FHAA does not apply to the sale or rental of a single family house by an owner as long as the owner does not own more than three single family houses at any one time and the house is sold or rented without the use of a real estate broker, agent or salesperson. Nor does it apply to a dwelling intended to be occupied by no more than four families, if the owner maintains and occupies one of the units as his or her residence. Subpart D of the regulations specifically address the prohibitions against discrimination because of disability and make the following practices unlawful:

- discrimination in the sale or rental, or to otherwise make unavailable or deny, a dwelling to any buyer or renter because of a disability of the buyer or renter, a person residing in or intending to reside in the dwelling or any person associated with the buyer or renter;
- discrimination against any person in the terms, conditions, or privileges of the sale or rental of a dwelling, or in the provision of services or facilities in connection with such dwelling; and
- inquiries to determine if a person has a disability or the severity of a disability.

A building need not comply with FHAA, if the last building permit is issued on or before June 15, 1990 and construction is completed without any permit renewals after this date. January 13, 1990 was the original deadline for issuance of a permit. The date was changed in the March 20, 1992 *Federal Register* to June 15, 1990 because this was the date HUD first published the proposed guidelines which provided detailed technical guidance on how to comply with FHAA's design and constructions requirements.

On March 6, 1991, HUD published the final Fair Housing Accessibility Guidelines in the *Federal Register*. The Guidelines are not mandatory, but simply provide technical guidance to builders and developers to assure a minimum level of accessibility. The guidelines reference ANSI A117.1-1986, Section 4.10 for elevators and Section 4.11 for wheelchair lifts.

1.5 AMERICANS WITH DISABILITIES ACT

President George Bush signed the Americans with Disabilities Act (ADA) on July 26, 1990. It is a sweeping law designed to extend to people with disabilities civil rights similar to those available on the basis of race, color, sex, national origin and religion through the Civil Rights Act of 1964.

Its purpose is to prohibit discrimination on the basis of disability in the private sector, and in state and local governments, particularly public accommodation and services, including transportation provided by public and private entities. It also includes provisions for telecommunications.

In a nutshell, the intent of ADA is to provide disabled people with accommodations and access equal to, or similar to, that available to the general public.

The ADA is modeled after the Civil Rights Act of 1964 and Title V of The Rehabilitation Act of 1973. The bill was originally drafted by the National Council on Disability who recognized that existing federal laws, which provided rights to federally related employment and accommodation, did not extend to public accommodations and commercial facilities that are not federally related. They contended that many state or local laws pertaining to access were nonexistent, obsolete or inadequate. Their proposed solution was a federal standard, which would apply to all buildings. The National Council on Disability began this process in 1986, which eventually lead to the signing of the ADA in July 1990.

Under Section 504 of the ADA, The Architectural and Transportation Barriers Compliance Board, also known as the "Access Board" or "ATBCB", was required to develop and issue guidelines, which would become the final ADA accessibility standards. The Access Board is an independent federal agency, which was originally founded to ensure that the requirements of federal disability laws were being met. The Board consists of 13 members of the general public appointed by the President, seven of whom must be disabled, as well as the heads of 12 government agencies including the Department of Education, DOD, GSA and USPS. Initially, the ATBCB was responsible for the MGRAD, which was the basis for UFAS until the ADA expanded their mandate to cover virtually all buildings in the U.S.

The Department of Justice (DOJ) is responsible for enforcement of certain sections of the ADA. It also was charged with development of final regulations to implement the provisions of Title II of the ADA, which pertains to public services (State and Local Government) and Title III of the ADA, which pertains to public accommodations. Section 306(c) of ADA requires that the DOJ include the Access Board's guidelines in its regulations. This resulted in the access guidelines, developed by the ATBCB, becoming the Federal Regulations enforced by the DOJ and DOT.

The Act itself is broken down into 5 elements or sections.

- **Title I – Employment**
Under the Act, no covered entity can discriminate against a qualified individual with a disability because of the disability. This relates to job application procedures, hiring, discharge, advancement, job training and other terms, conditions and privileges of employment.

- **Title II - Public Services**
The Title II regulation covers "public entities." "Public entities" include any state or local government and any of its departments, agencies or other instrumentalities. All activities, services and programs of public entities are covered, including activities of state legislatures and courts, jails, town meetings, police and fire departments, motor vehicle licensing and employment.

- Unlike Section 504 of Rehabilitation Act of 1973, which only covers programs receiving federal financial assistance, Title II extends to all the activities of state and local governments whether or not they receive federal funds.

This section also is aimed at improving access to bus and rail public transportation operated by state and local governments. Individuals with disabilities cannot be denied the benefit of these and other public services.

Public entities must ensure that newly constructed buildings and facilities are free of architectural and communication barriers that restrict access or use by individuals with disabilities. When a public entity undertakes alterations to an existing building, it must also ensure that the altered portions are accessible. The ADA does not necessarily require retrofitting of existing buildings to eliminate barriers, (moving services and activities to accessible locations to provide "program accessibility" may be an alternative) but it does establish a high standard of accessibility for new buildings.

- Public entities may choose between two technical standards for accessible design: The Uniform Federal Accessibility Standard established under the Architectural Barriers Act, or the Americans with Disability Act Accessibility Guidelines, adopted by the Department of Justice for places of public accommodation and commercial facilities covered by Title III of the ADA. They cannot pick and choose requirements from either standard. They must pick either ADAAG or UFAS and comply with all the provisions in the selected standard.
- The elevator exemption (see Chapter 2, Section 2.7) for small buildings under ADAAG would not apply to public entities covered by Title II.

- **Title III - Public Accommodation**
    Title III prohibits denial of the full and equal enjoyment of goods, services, facilities and accommodation. It applies specifically to public accommodation; businesses and services offered by private entities, such as hotels, restaurants and shops. Title III also applies to privately owned places of employment.

    Public accommodation is defined as including the following:
    1. Hotels or places of lodging
    2. Restaurants or bars
    3. Theaters or stadiums
    4. Auditoriums or convention halls
    5. Any retail shop
    6. Service establishments such as banks, travel agents or laundromats
    7. Public transportation depots
    8. Museums, libraries or galleries
    9. Zoos or amusement parks

10. Any place of education
11. Day care centers or public shelters
12. Recreation facilities like a bowling alley

Title III also includes commercial facilities, which are defined as office buildings, factories and other places in which employment will occur.

Of significance to the elevator industry is the fact that Title III establishes very specific accessibility requirements for new construction and alterations for all of these categories of buildings, including design requirements for wheelchair lifts and elevators.

Title III typically does not apply to churches, private clubs and residential buildings. If a religious entity or private club allows their facility to be used for outside activities, the facility may then be subject to the requirements in Title III. The definition of a religious entity is:

> A religious organization or an entity controlled by a religious organization, including a place of worship.

The exemption from Title III includes all of the activities of a religious entity whether religious or secular. The exemption is intended to have broad application. For example, a parochial school that is sponsored by a religious order would be exempt, even if it has a lay board of directors. Even though the religious entity is operating a facility that would otherwise be considered a place of public accommodation, its operations are exempt from the Title III requirements. However, if the religious entity, rents all or a portion of its facilities to a private entity to operate a place of public accommodation, then the private entity is subject to the Title III requirements. The religious entity is still exempt. The obligations of a landlord for a place of public accommodations do not apply if the landlord is a religious entity.

A nonreligious entity running a place of public accommodation in space donated by a religious entity is exempt from Title III requirements. The nonreligious tenant is subject to Title III only if the space is rented. Title III also does not apply to a "private club." An entity is a private club for purposes of the ADA if it is a private club under Title II of the Civil Rights Act of 1964, which prohibits discrimination on the basis of race, color and national origin by public accommodations. Courts have been most inclined to find private club status in cases where:

1. Members exercise a high degree of control over club operations.
2. The membership selection process is highly selective.
3. Substantial membership fees are charged.
4. The entity is operated on a nonprofit basis.
5. The club was not founded specifically to avoid compliance with Federal civil rights laws.

A private club loses its exemption when it makes its facilities available to nonmembers as places of public accommodation.

A residential facility, which typically is covered by the requirements under the Fair Housing Amendments (FHAA) Act 9, (see Chapter 1, Section 1.4) may also be subject to the requirements of Title III of ADA. The analysis for determining whether a facility is covered by the ADA is entirely separate and independent from the analysis used to determine coverage under the FHAA. A facility can be a residential dwelling under the FHAA and still fall in whole or in part under at least one of the 12 categories of places of public accommodation. An example would be an apartment building, which leases space to a health care provider or has a rental office on the premises.

- **Title IV - Telecommunications**

  This section ensures that interstate telecommunication service is available to hearing-impaired and speech-impaired individuals.

- **Title V – Miscellaneous**

  Title V covers explanations, exemptions, directives and mandated studies.

Under the ADA the ATBCB was charged with preparing guideline regulations. Those guidelines were utilized by the DOJ in setting the ADA regulations.

Compliance with ADAAG must be met as a minimum where Title III is applicable. Local regulations can be more stringent. Where less stringent, the ADAAG provisions take precedent. The requirements in ADAAG are based on the provisions in the 1980 version of the ANSI A117.1. If you obtain a copy of the guidelines, it will be easy to note the difference as they are highlighted by italics text. The ADAAG regulations are covered in detail in Chapter 4.

1.6 ADAAG - THE FUTURE

The DOJ's Title II regulations give state and local governments the option of choosing between designing, constructing or altering their facilities in conformance with UFAS or with ADAAG, except that if ADAAG is chosen, the elevator exemption (see Chapter 2, Section 2.7) contained in Title III of the ADA does not apply.

When the DOJ published its Title III regulations, it noted that the Access Board would be supplementing ADAAG in the future to include additional guidelines for state and local government facilities. The DOJ further stated that it anticipated that it would amend its Title II regulations to adopt ADAAG as the accessibility standards for state and local government facilities after the Access Board supplemented ADAAG. Adopting essentially the same accessibility standards for Title II and III of the ADA will ensure consistency and uniformity of design in the public and private sectors throughout the country.

ADA design standards are developed in a two-step process. They are first published by the Access Board as a minimum guideline for the DOJ and DOT. The DOJ and DOT are responsible for adopting enforceable standards that are consistent with the minimum guidelines. On January 13, 1998, the Access Board published in the *Federal Register* amendments to ADAAG that establish guidelines for state and local government facilities and building elements designed for children's use. The amendments change several sections of ADAAG and add new sections that cover access to Title II, judicial, legislative and regulatory facilities (section 11) and to detention and correctional facilities (section 12), and provide alternate specifications based on children's dimensions for various building elements such as water closets and lavatories. The DOJ, which has jurisdiction over the mentioned Title II facilities, has not yet adopted these amendments as part of the enforceable standard under ADA.

To further the goal of uniform standards, the Access Board intends to use ADAAG as the accessibility guidelines for federally financed facilities covered by the Architectural Barriers Act of 1968. Under Section 502 of the Rehabilitation Act of 1973, the Access Board is responsible for establishing guidelines for accessibility standards issued by other federal agencies pursuant to the Architectural Act of 1968. The Access Board has publicly stated that the next edition of ADAAG will include provisions that are appropriate for federal buildings (e.g., post offices, military facilities) that are to be used in place of its current guidelines for federally financed facilities. Standards issued by other federal agencies pursuant to the Architectural Barriers Act must be consistent with the Board's guidelines. Those federal agencies responsible for issuing accessibility standards under the Architectural Barriers Act will initiate separate rulemaking to adopt standards consistent with ADAAG as supplemented in place of UFAS.

In 1994, the Access Board began a review of ADAAG. An ADAAG Review Advisory Committee was appointed. The April 6, 1994, *Federal Register* notice summarized the objectives of the Advisory Committee:

> "The Architectural and Transportation Barriers Compliance Board (Access Board) announces its intent to establish an advisory committee to review the Americans with Disabilities Act Accessibility Guidelines (ADAAG) for buildings and facilities and requests applications from interested organizations for members to serve on the committee. The committee will make recommendations to the Access Board for updating ADAAG to ensure that the guidelines remain consistent with technological developments and changes in national standards and model codes, and meet the needs of individuals with disabilities."

The ADAAG Review Federal Advisory Committee issued a final report with recommendations for a new ADAAG September 30, 1996. The Access Board accepted the report and at its March 1999 meeting the Board unanimously approved a rule to completely revise and update its ADAAG. This moves the rule one step closer to publication. Once published, the rule will be available for public comment. This will be the first comprehensive update of the guidelines since they were originally issued in July 1991.

The Board adopted the complete contents of the proposed rule. This includes not only the text of the rule providing updated scoping and technical requirements, which the Board previously approved last September, but also new illustrations and advisory material (commentary) developed in-house by Board staff, and an accompanying discussion of the changes known as the "preamble" in the published rule. The preamble will provide a section-by-section description of the changes and will ask questions of the public concerning various issues and provisions. Public comment, including the information and input provided in response to questions, greatly assists the Board in finalizing a proposed rule.

So when does the proposed rule hit the streets? Not until a few regulatory actions are completed beforehand. First, the proposed rule, along with a regulatory assessment, must be reviewed by the Office of Management and Budget (OMB), which reviews most federal regulations. Once cleared by OMB, the rule will be published in the *Federal Register* and be made available for public comment.

The Board intends to provide a 120-day comment period. During this time, it will hold several public hearings that will provide a forum for submitting comment. The date and places for these hearings will be indicated in the proposed rule along with instruction for submitting comments to the docket. The text of the rule will contain a new ADAAG based on recommendations from the ADAAG Review Advisory Committee. The committee's report is available from the Board (publication S29) and its web site (www.access-board.gov). In addition, the rule will provide updated guidelines for federally funded facilities under the Architectural Barriers Act (ABA). The ABA guidelines will be modeled after the new ADAAG so that a consistent level of access is required for federal facilities and for facilities covered by the ADA. It is anticipated the proposed revised ADAAG will be published in the late fall of 1999. Once the public comment period is closed, the Access Board will review the public comments, draft responses and revise the draft where appropriate. At some undetermined future date, a revised ADAAG will be published in the *Federal Register*. Before the revised ADAAG becomes effective, they must be incorporated into the DOJ and DOT regulations. A similar procedure must be followed by DOJ and DOT which can be done simultaneously or after the revised ADAAG is approved by the Access Board.

1.7 LIMITED USE/LIMITED APPLICATION ELEVATORS

The ASME A17.1 Safety Code for Elevators and Escalators is often referred to as a living document. As technology advances, the ASME A17.1 Code adopts requirements that addresses the safety concerns associated with new technology.

In the late 1980s, the ASME A17 Main Committee became aware that a need existed for a small elevator similar to a private residence elevator (ASME A17.1, Part V) except suitably designed for public use.. An Ad Hoc Committee was appointed in June 1989 to research the feasibility of ASME A17 responding to this need. The Ad Hoc Committee's report states in part:

"......Today, we are again seeing a need expressed for a new class of limited-use/limited-access elevator one which is commonly installed in an existing building such as a church or a lodge, is not used extensively, and is necessary to provide access to persons who cannot negotiate stairs. From a code and regulatory standpoint, this has been addressed in two ways: jurisdictions have granted variances to allow the use of private residence type elevators, or they have developed their own standards.

"The initial task of the ad hoc committee was to develop parameters for the equipment which would limit its use and its access, thus allowing the modification of safety requirements.

"It should be noted that the Committee did not include accessibility issues amongst these parameters and, in fact, equipment which meets these limitations would not conform with the specifications in the A117.1 standard. It was determined that the requirements for size and operating characteristics included in accessibility standards necessitate the use of equipment which would have to conform to Parts I, II and III of the A17.1 Code. Also, the marketplace has shown that conformance with accessibility standards is not a prerequisite for this type of equipment.

"The development of alternative requirements for this type of equipment is bound to be seen as an effort to create a cheap elevator with a lower standard of safety - this is not the case. For example, special purpose personnel elevators have less stringent requirements than passenger elevators, but they are no less safe since the differences in the requirements are justified based on differences in application. The Committee believes that, likewise, we will be able to develop alternative requirements for this type of equipment based on its limited use and limited access; and that we will be able to accomplish this without compromising safety.

"A preliminary review of Parts I, II and III of the Code has indicated requirements which can be modified. There are other requirements that actually need no revision since they already allow alternatives for slower or smaller equipment. The following are some examples:

- Pits and overhead clearances could be substantially reduced, provided that other provisions are made to ensure the safety of persons working underneath and on top of the car.
- Machine rooms could be reduced in size, and/or eliminated based on the use of small size equipment.
- Spring buffers or bumpers could be used instead of oil buffers due to slower speeds.
- Smaller ropes, machines, plungers, cylinders, etc. could be used due to the limitations on size and lower usage.

- Firefighters' service would not be require due to the limitation on travel.

CONCLUSIONS AND RECOMMENDATIONS

The Committee has concluded that: (1) a need has been shown for a limited-use/limited-access elevator; and (2) it is possible to develop alternative requirements without compromising safety.

The ASME A17 Main Committee accepted the Ad Hoc Committee report in April 1991 and the A17 Limited-Use/Limited-Access, later changed to Limited-Use/Limited-Application Committee (ASME A17 LU/LA Committee) was appointed. Brian Black of the Eastern Paralyzed Veterans Association was appointed chairman of the Committee. He immediately proceeded to canvass all interested parties to solicit Committee Members. The first meeting of the ASME A17 LU/LA Committee was held in September 1991.

In the period between the ASME A17 Main Committee acceptance of the Ad Hoc Committee report and the first meeting of the ASME A17 LU/LA Committee, the ADAAG regulations were published. It became clearly evident that one of the basic assumptions made by the Ad Hoc Committee was no longer valid. ADAAG § 4.1.3(5) Exception 1 states that if an elevator is provided, even when not required, the elevator shall comply with § 4.10. Clearly, a LU/LA elevator complying with ADAAG § 4.10 was not envisioned by the Ad Hoc Committee. The ASME A17 LU/LA Committee was in a quandary. While accessibility is not within the scope of the ASME A17.1 Code, there was not much point in proceeding if a LU/LA elevator would not be acceptable under ADA.

In the summer of 1992, after considerable discussion among Committee Members, the ASME A17 LU/LA Committee met with the ATBCB Technical Staff. The position put forth by the ASME A17 LU/LA Committee was that the ADAAG requirement that all elevators comply with § 4.10 would discourage the installation of an elevator when they are not required by ADA. The Committee's position was that a LU/LA elevator, while not providing the same degree of accessibility, an elevator complying with ADAAG § 4.10, it would nevertheless, provide an economical, accessible means of access to upper floors. If the current provisions are not changed, upper floors would typically be inaccessible since an elevator complying with ADAAG § 4.10 would not be economically feasible.

The ATBCB staff encouraged the ASME A17 LU/LA Committee to continue. They indicated that once ASME A17 recognized a LU/LA elevator, they would be amenable to considering accessibility requirements for them. The same was true for the CABO/ANSI A117.1 Standard. The ASME A17 Committee is responsible for "safety" while ANSI A117 is responsible for "accessibility." The ASME A17 LU/LA requirement will recognize operations, features, etc. that are considered safe. Conversely, the ANSI A117.1

Standard may not allow some of these operations, features, etc. that do not provide accessibility.

The ASME A17 Committee was encouraged to proceed with the development of safety requirements for LU/LA elevators. ASME A17.1-1993, Addend A17.1b-1995 recognized Limited-Use/Limited-Application elevators for the first time. NEII then took the initiative to draft accessibility requirements for LU/LA elevators and submitted the proposal to the CABO/ANSI A117 Committee and ADAAG Review Advisory Committee. The final report of the ADAAG Review Advisory Committee and ICC/ANSI A117.1-1998 includes requirements for an accessible LU/LA elevator.

A final caution regarding LU/LA elevators. There are a number of states that currently recognize LU/LA elevators. At the meeting between the ASME A17 LU/LA Committee and the staff of ATBCB, a chief elevator inspector for one of the states poised the following question:

"My state recognizes a LU/LA elevator. It does not comply with all the provisions in ADAAG § 4.10, as an example, car size and automatic door operation. Is this elevator permitted under ADA?"

The ATBCB Staff response was as follows:

"If the elevator is installed in a building subject to the provision of ADAAG, the elevator must comply with all the provision in § 4.10."

One can only conclude, that in a building subject to ADA regulations (see Chapter 3) the installation of a LU/LA Elevator is currently unacceptable. If the building is not subject to ADA regulations but subject to ICC/ANSI A117.1-1998, then a LU/LA elevator can be provided for accessibility. A building owner who installs a LU/LA elevator would be subject to the enforcement provision of the Act, including civil penalties and reimbursement of a prevailing parties attorney's fees.

1.8 EVACUATION OF DISABLED IN A FIRE

The NEII Handicapped Standard has since the first edition in 1976 stated:

"Elevators cannot be considered as exits in an emergency. Consideration should be given to emergency evacuation. A definite plan is required to assist the physically handicapped, particularly those in wheelchairs. Elevators may not be available during a fire."

I have attended numerous meetings with disability organizations where the elevator industry raised this concern. The typical response during the 1970s and early 1980s was don't worry about us in a fire; we will take our chances, just make the building accessible.

In recent years, many organizations have been wrestling with this concern, including the model building codes and NFPA Life Safety Code. The Access Board also was concerned and in formulating the regulations they mandated for new construction, ADAAG § 4.1.3(9), added that exits be accessible or areas of rescue assistance be provided. Typically, floors above ground level are exited by stairways. Stairways are not considered accessible, thus areas of rescue assistance must be provided. An area of rescue assistance, ADAAG § 4.3.11, is a location where the disabled can, in theory, safely await evacuation in a fire. Evacuation of the upper floors would then proceed using Phase II Firefighters' Service.

A number of firefighting and fire protection engineers are questioning the effectiveness of the areas of rescue assistance, also known as refuge areas. Some feel that areas of rescue assistance not located adjacent to elevators are a potential problem. They are advocating making the elevator lobby the only recognized area of rescue assistance. They further recommend pressurizing the elevator hoistway and lobby to keep these areas smoke free.

Others are voicing a need to automatically program elevators to evacuate occupants, able bodied and disabled, from the fire floor and floors surrounding it. In 1992, the National Institute of Standards and Technology (NIST) hosted a workshop on this approach. (See Appendix A for purchasing information on workshop paper that has been published by NIST.) No conclusions were reached, but it was apparent that numerous obstacles must be overcome. In order to provide continuous reliable elevator service in a fire, the environment an elevator operates in must be protected. Key concerns include, reliable building power, smoke free elevator lobbies, hoistways and machine rooms, tolerable ambient temperature in machine rooms and, last but not least, keeping water out of elevator hoistways. These issues are continuing to be debated by the fire protection community. The elevator industry must contribute to the discussions or suffer the results of regulations imposed by organizations that do not understand elevator safety and technology.

OVERVIEW OF ADA TITLE III REGULATIONS

2.

2. OVERVIEW OF ADA TITLE III REGULATIONS

2.1 ATBCB's ACCESSIBILITY GUIDELINES

The ADA stipulated that the U.S. ATBCB, also known as the Access Board, develop guidelines (standards) for existing buildings, alterations and new construction.

The Access Board released proposed ADA guidelines for public comment on January 22, 1991. They heard testimony from over 450 individuals and groups and received over 12,000 pages of written comment. The final guidelines were published in the *Federal Register* July 26, 1991, and began to go into effect January 26, 1992.

The standard is the Americans with Disabilities Act Accessibility Guidelines for Buildings and Facilities, or ADAAG. It is a very comprehensive standard prepared by the Federal Access Board to ensure that the buildings covered by Title III are readily accessible to and usable by individuals with disabilities in terms of architecture, design and communication. These guidelines have been incorporated as an appendix to the DOJ, ADA regulations.

ADAAG is modeled after the Uniform Federal Accessibility Standards (UFAS) and is generally consistent with the Federal Minimum Guidelines and Requirements for Accessible Design (MGRAD). The format and numbering follow the ANSI A117.1 - 1980 Standard.

2.2 GENERAL

Public accommodations must remove architectural barriers and communication barriers that are structural in nature in existing facilities, when it is readily achievable to do so. Architectural barriers are physical elements of a facility that impede access by people with disabilities. Those barriers include more than the obvious impediments such as steps that may prevent wheelchair access. In many facilities, telephone, drinking fountains, elevator controls, etc. are mounted at heights that prevent access by people who use wheelchairs.

Communication barriers that are structural in nature are barriers that are an integral part of the physical structure of a facility. Examples include conventional signage, such as elevator car control markings, which cannot be read by people who have vision impairments or an audible emergency alarm, which cannot be heard by people with hearing impairments.

According to the Act, readily accessible and usable does not mean total accessibility, but rather that a building or a part of a building be provided with a reasonable degree of accessibility to accommodate disabled people.

Accessibility, in general, means being in full compliance with ADAAG such that the site and building can be approached, entered and used by individuals with disabilities, including those affecting mobility, sensory or cognitive functions. In other words, accessibility also applies to individuals who are deaf or blind.

The rules vary for existing facilities, alterations and new construction. In existing facilities where retrofitting may be expensive, the access requirements are less stringent than for new construction and alterations where accessibility can be incorporated in the design and construction without a significant cost increase.

2.3 EXISTING FACILITIES

Starting with existing facilities, ADA requires that architectural and communication barriers be removed immediately, providing that it is readily achievable to do so. Readily achievable means easily accomplishable and able to be carried out without much difficulty or expense. There is no formula in the regulations but factors to be considered include cost, financial resources of the facility and the financial resources of the parent company.

Determining what is readily achievable is a case-by-case judgment. Factors to consider are:

- The nature and cost of the action;
- The overall financial resources of the site or sites involved; the number of persons employed at the site; the effect on expenses and resources; legitimate safety requirements necessary for safe operation, including crime prevention measures; or any other impact of the action on the operation of the site;
- The geographic location, and the administrative or fiscal relationship of the site or sites in question to any parent corporation or entity;
- If applicable, the overall financial resources of any parent corporation or entity; the overall size of the parent corporation or entity with respect to the number of its employees; the number, type and location of its facilities; and
- If applicable, the type of operation or operations of any parent corporation or entity, including the composition, structure and functions of the work force of the parent corporation or entity.

If the public accommodation is in a facility that is owned or operated by a parent entity, that conducts operations at many different sites, the public accommodation must consider the resources of both the local facility and the parent entity to determine if removal of a particular barrier is "ready achievable." The administrative and fiscal relationship between the local facility and the parent entity must also be considered in evaluating what resources are available for any particular act of barrier removal.

It is clear that the ADA does not require major expenditures, such as the installation of an elevator, in existing facilities. A building generally would not be required to remove a barrier to physical access posed by a flight of stairs by installing an elevator. The readily achievable standard does not require barrier removal that requires extensive restructuring or major expense.

Thus, it should be obvious there is no definitive answer to what is readily achievable because determination as to what barriers can be removed without much difficulty or expense can only be made on a case-by-case basis.

The ADA establishes priorities for barrier removal. There is no use having an accessible washroom if you can't get into the building. The DOJ regulation recommends priorities for removing barriers in existing facilities. Because the resources available for barrier removal may not be adequate to remove all existing barriers at any given time, the regulation suggests a way to determine which barriers should be made less severe or eliminated first. The purpose of these priorities is to encourage long-term business planning and to maximize the degree of barrier removal that will result from the expenditure. These priorities are not mandatory. Public accommodations are free to exercise discretion in determining the most effective barrier removal measures to undertake in their facilities. The priorities are:

- Access onto site and into building;
- An accessible washroom;
- Access to goods and services; and
- Access to remaining areas, goods, services or accommodations.

This final priority may include access to something other than an "area," such as installing volume controls on a pay phone.

There is no requirement to undertake major alterations to make facilities accessible. However, if alterations are undertaken, the altered area must be brought up to the ADAAG requirements and made accessible to the maximum extent feasible. To the maximum extent feasible applies to the occasional situation where the nature of the facility makes it virtually impossible to comply with the guidelines. Occasionally, it may be impossible to comply with all the requirements in ADAAG. In such a case, only those items technically feasible would have to be made accessible. The fact that compliance with ADAAG increases costs does not mean compliance is technically infeasible. Cost is not a consideration. Even when it is technically infeasible to comply with the requirements for individuals with certain disabilities such as wheelchair users, the alteration must still comply with the requirements for individuals with other impairments.

An alteration is any change that affects usability under the ADA, alterations are defined as:

1. Remodeling
2. Renovation
3. Rehabilitation

4. Historic renovation
5. Changes in structural elements
6. Other extraordinary repairs

Normal maintenance, painting or wallpapering does not constitute an alteration unless they affect usability.

An interesting requirement is that if there is an alteration to a primary function area, an accessible path of travel to that area must be provided. A path of travel is an unobstructed route from outside the building to the primary function area, and includes restrooms, telephone and drinking fountains in the area. This means that an alteration to an upper level of a building could trigger the requirement for an elevator or a wheelchair lift to create an accessible path of travel. Alterations to provide an accessible path of travel are required whenever it is not disproportionate to the cost of the original alteration. The cost to create the path of travel does not have to exceed 20% of the alteration cost of the primary function area. A primary function area is defined as any room or space where the major activities of the facility are carried out. It includes offices and work areas in commercial facilities. In public accommodations it includes both customer service areas and work areas. An elevator machine room would not be considered a primary function area.

There are some alterations that will never trigger the path of travel requirement. The DOJ regulation states that alterations to windows, hardware, controls, electrical outlets and signs do not trigger path of travel requirements. If they affect usability, however, they are still considered alternations and must be done in an accessible fashion. The path of travel requirement is not triggered if alteration work is limited solely to the electrical, mechanical, or plumbing system, hazardous material abatement, or automatic sprinkler system retrofitting, unless the project involves alteration to elements required to be accessible. Finally, the path of travel requirement is not triggered where a public accommodation is undertaking alterations only to meet its obligation to remove barriers. Thus, the installation of elevator car control and elevator door jamb markings would not trigger the path of travel requirements.

A building cannot evade the path of travel requirement by doing several small alterations each of which, if considered by itself, would be so inexpensive that adding 20% would not result in addition of any path of travel features. Whenever an area containing a primary function is altered, other alterations to that area or to other areas on the same path of travel, made within the preceding three years are considered together. In other words, all the alterations to a primary function area made within the preceding three years are considered in calculating the 20% required to be spent on the accessible path of travel. Only alterations after January 26, 1992 are counted.

2.4 NEW CONSTRUCTION

The rules for new construction are straightforward. The ADAAG regulations apply to the entire facility as required by Subpart D of the DOJ Title III regulations, unless structurally impracticable. A brief idea of the scope of the guidelines for alterations and new construction, include requirements for such things as fire alarms, areas of refuge, doors, entrances, washrooms, signs, telephones, fountains, ramps, stairs, parking and, of course, elevators and wheelchair lifts.

Departures are permitted from particular requirements where alternative designs and/or technologies provide substantially equivalent or greater access to and usability of the facility. ADAAG refers to this as "equivalent facilitation." Unlike typical code enforcement there is no authority having jurisdiction to apply for a variance or to obtain approval before proceeding. The ADA regulations leave it up to the owner/operator to justify equivalent facilitation has been provided, when challenged.

The phrase "structurally impracticable" means that unique characteristics of the land prevent the incorporation of accessibility features in a facility. In such a case, the new construction requirements apply, except where the entity can demonstrate that it is structurally impracticable to meet those requirements. This exception is very narrow and should not be used in cases of merely hilly terrain. The DOJ has publicly stated that they expect the exception will be used in only rare and unusual circumstances. Even in those circumstances where the exception applies, portions of a facility that can be made accessible must still be made accessible. In addition, access is required to be provided for individuals with other types of disabilities, even if it were structurally impracticable to provide access to individuals who use wheelchairs.

2.5 EFFECTIVE DATES

The effective date for existing buildings and alterations was January 26, 1992. Small businesses had up to an additional year to comply.

New buildings designed and constructed for first occupancy after January 26, 1993 are required to comply.

The term designed and constructed for first occupancy after January 26, 1993 is fulfilled by two criteria:

1. If the last application for a building permit or permit extension is filed (and, where applicable, certified as complete by the local jurisdiction) after January 26, 1992; and
2. If the first certificate of occupancy is issued after January 26, 1993.

The date of January 26, 1992 is relevant only to the last application for a permit or permit extension for a facility. Thus, if an entity has applied for only a foundation permit, the date of that application has no effect, because the entity must also apply for and receive a permit later for the actual superstructure.

2.6 ELEVATORS REQUIRED

Vertical transportation to multistory buildings is specifically covered in the ADA and ADAAG. ADAAG states that one passenger elevator complying with Section 4.10 shall serve each landing, including mezzanines in all multistory buildings and facilities, unless exempted. If more than one elevator is provided, each full passenger elevator must comply with the ADAAG requirements.

Exemption 1 states that elevators are not required in facilities that are less than 3 stories or that have less than 3,000 square feet per story unless the building is a shopping center, shopping mall, or the professional office of a health care provider, or another type of facilities as determined by the Attorney General. The Attorney General (DOJ regulation) has determined that the public portions of transportation facilities (train depots, bus stations, airports) are not exempt. The regulations further state that, in new construction, if a building or facility is eligible for this exemption and a full passenger elevator is planned, that elevator shall meet all the requirements of ADAAG and shall serve each level in the building. A full passenger elevator that provides service from a garage to only one level of a building or facility is not required to serve other levels.

Exception 2 states that an elevator is not required for elevator pits, elevator penthouses, mechanical rooms, and piping or equipment catwalks.

Exception 3 states that accessible ramps complying with the regulations may be used in lieu of elevators.

Exception 4 states that wheelchair lifts complying with Section 4.11 of ADAAG and applicable state or local codes may be used in lieu of an elevator only under the following conditions:

1 - To provide an accessible route to performing areas in an assembly occupancy.
2 - To comply with the wheelchair viewing position line of sight and dispersion requirements of ADAAG.
3 - To provide access to incidental occupiable spaces and rooms which are not open to the general public and which house no more than 5 persons, including but not limited to equipment control rooms and projection booths.
4 - To provide access where existing site constraint or other constraints make use of ramps or elevators infeasible.

2.7 ELEVATORS NOT REQUIRED

One of the more interesting exemptions of ADA, presumably a cost saving concession to building owners, is the fact that elevators are not required in new or altered facilities of less than 3 stories, or if the building has less than 3,000 square feet per floor. The ground floor will have to be accessible, and the other floors will have to be in compliance with the guidelines, except no elevator is required. A story is occupiable space, designed for human occupancy and equipped with one or more means of egress, light and ventilation. Basements designed or intended for occupancy are considered stories. Mezzanines are not considered stories, as they are levels within stories.

The exception does not apply to shopping malls or offices of health care providers. Those facilities require elevators. It should be noted that this exception might well be superseded by the requirements of local building codes and access regulations that, if more stringent, would apply. A "shopping center or mall" is either:

- A building with five or more "sales or retail establishments," or
- A series of buildings on a common site, either under common ownership or common control or developed together, with five or more sales or retail establishments.

Included within the phrase sales and retail establishments are those types of stores listed in the fifth category of places of public accommodations, i.e., bakery, grocery store, clothing store, hardware store, etc. (See Chapter 1, Section 1.5). The term includes floor levels containing at least one such establishment, or any floor that was designed or intended for use by at least one such establishment. The definition of shopping center or mall is slightly different for purposes of alterations. A shopping center or mall is defined in the alterations provisions as a series of existing buildings on a common site connected by a common pedestrian route above or below the ground floor. This definition was included to avoid requiring several separate elevators in buildings that were initially designed and built independently of one another. The common pedestrian route would allow access to all the stores to be provided by a single elevator.

A professional office of a health care provider is defined as a location where a state regulated professional provides physical or mental health services to the public. Factors that DOJ will use to determine whether a facility was designed or intended for occupancy by a health car provider include:

- whether the facility has special plumbing, electrical, or other features needed by health care providers;
- whether the facility was marketed as a medical office center; and
- whether any of the establishments that actually first occupied the floor were, in fact, health care providers.

2.8 ENFORCEMENT AND LEGAL REMEDIES

It is important to recognize that the ADA regulations are not codes or standards in the conventional sense. ADA is a federal civil rights law which state and local inspectors are not required or allowed to enforce. Even then, they do not have the right to interpret or give a variance to ADAAG.

ADA will be enforced and interpreted, as other civil rights laws are, by the action of an aggrieved party through the judicial system, in other words, when someone files a complaint. Private individuals who are being subject to discrimination or believe they are about to be discriminated against, may bring lawsuits in which they can obtain court orders to stop discrimination. Individuals may also file complaints with the DOJ. The DOJ will investigate the complaint and conduct compliance review. Complaints may be sent to the following address:

> Office of the Americans with Disabilities Act
> Civil Right Division
> United States Department of Justice
> P. O. Box 66738
> Washington, DC 20035-6738
> (800) 669-4000
> www.usdoj.gov/crt/ada/adahom1.htm

The Attorney General may bring a civil action in any appropriate district court, if he has reasonable cause to believe any person or group is engaged in a pattern or practice of discrimination or any person or group has been discriminated against and the discrimination raises an issue of general public importance. In these cases, the Attorney General may seek monetary damages for the individual victim and civil penalties to vindicate the public interest.

In a private suit at the request of the plaintiff or defendant, and if the court permits it, DOJ can intervene in the civil action, if it determines that the case is of general public importance. The court may also appoint an attorney for the plaintiff and may permit him or her to commence the civil action without first paying fees, costs, or security.

Remedies available in a private suit may include a permanent or temporary injunction, restraining order, or other order, but not compensatory or punitive money damages or civil penalties. In the case of violations of the requirements for readily achievable barrier removal or for accessible new construction and alterations, remedies to correct a violation may, as appropriate, include an order to alter the facilities that do not meet the requirements of the ADA to make them readily accessible to and usable by individuals with disabilities. Also, the remedies may include requiring the provision of an auxiliary aid or service, modification of a policy, or provision of alternative methods of barrier removal.

The remedies available in a civil suit brought by DOJ include those available in an action brought by an individual, such as an order granting temporary, preliminary, or permanent relief; requiring that facilities be made readily accessible to and usable by individuals with disabilities; requiring provision of an auxiliary aid or service; or modification of a policy, practice, or procedure. In addition, the court may award other appropriate relief, including, if requested by DOJ, monetary damages to individual victims of discrimination. Monetary damages do not include punitive damages. They do include, however, all forms of compensatory damages, including out-of-pocket expenses and damages for pain and suffering.

Also, to vindicate the public interest, the court may assess a civil penalty against the covered entity in an amount:

- Not exceeding $50,000 for a first violation; and
- Not exceeding $100,000 for any subsequent violation.

The prevailing party (other than the U.S.) in any action or administrative proceeding under ADA may recover attorney's fees in addition to any other relief granted. The prevailing party is the party that is successful and may be either the complainant (plaintiff) or the covered entity against which the action is brought (defendant). The defendant, however, may not recover attorney's fees unless the court finds that the plaintiff's action was frivolous, unreasonable, or without foundation, although it does not have to find that the action was brought in subjective bad faith. Attorney's fees include litigation expenses, such as expert witness fees, travel expenses, and costs. The U.S. is liable for attorney's fees in the same manner as any other party, but is not entitled to them when it is the prevailing party.

The ADA encourages the use of alternative means of dispute resolution, including settlement negotiations, conciliation, facilitation, mediation, fact-finding, mini trials and arbitration to resolve disputes, where appropriate and to the extent authorized by law. In appropriate cases, these types of procedures may be faster, more efficient and less expensive than the judicial and administrative procedures available under the ADA. Alternative means of dispute resolution, however, are intended to supplement the procedures provided in the ADA, not to replace them. Use of alternative procedures is completely voluntary and must be agreed to by the parties involved.

There is no provision for state or local civil rights agencies to directly enforce ADA. They can, however, enforce state or local laws that incorporate the standards of the ADA, or they can set up alternative dispute resolution mechanisms.

The major concern of building owners and tenants is the huge anticipated cost of implementing the ADAAG regulations. It is estimated that annual construction costs related to ADA will exceed two hundred million dollars. Many building owners are very concerned about carrying the financial burden alone, with minimal government assistance. There are also concerns about the technical aspects of compliance, and the need for expert technical assistance. This concern has become very apparent as the regulations are not always clear and no government agency is yet willing to interpret ADAAG. There is a need for government programs to help solve the technical problems of compliance. A final concern, which is shared by regulatory officials, is the enforcement of ADA through the legal system. Currently, the only official method of obtaining interpretations is through the federal courts. The courts are not set up to interpret building code issues.

For code enforcement authorities, the ADA throws a few curve balls into a fairly well established system. The first is the introduction of yet another standard (ADAAG), which contains requirements, terms and concepts that are not always clear and are not always consistent with the codes that are presently being used. As previously stated, the courts will be responsible for final interpretations and settlement of the resulting disputes. There is concern that the courts are not equipped in terms of capacity or expertise to deal with code interpretation and enforcement.

Another issue is the role of DOJ in physical site inspections to determine in response to the filing of a complaint, that a building owner is actually in violation of the ADAAG requirements, whether the requirements are readily achievable, and once a correction is court ordered, to inspect for completion and compliance. Clearly, the DOJ is not a federal building inspection agency.

A third concern is the process of enforcement itself. Traditionally, enforcement authorities, such as elevator inspectors, are able to identify code violations and, empowered by local legislation, are able to obtain compliance without total reliance on the judicial system. They are also, as a rule, empowered to make interpretations locally, giving flexibility to compliance with the intent of the code. The DOJ enforcement of ADAAG, on the other hand, is a reactive form of enforcement that requires no initial inspection for compliance and relies on individuals to file legal complaints to bring buildings into compliance.

Compliance then becomes highly interpretive and is almost voluntary for the building owner, until a complaint or suit is filed. For building owners who seek voluntary compliance, there is no inspection mechanism to confirm that compliance has been achieved.

2.9 STATE AND LOCAL ACCESSIBILITY REQUIREMENTS

The DOJ is well aware of the enforcement problems and its likely ineffectiveness as an inspection organization. The solution in the eyes of DOJ is for state or local governments to apply to the Attorney General, who in consultation with the ATBCB, may certify that state or local codes meet or exceed the ADAAG requirements.

The ADA authorizes the Attorney General to certify that state and local building codes, or similar requirements meet or exceed ADAAG. Certification offers many potential advantages including:

- When designing, constructing, or altering a building in accordance with an applicable state or local code that has been certified by DOJ, the designer or contractor will need to consult only that one code, in order to determine the applicable Federal, State, and local requirements.
- There will be some degree of assurance in advance of construction or alteration that the ADA requirements will be met.
- Subject to a lawsuit, compliance with a certified code will be rebuttable evidence of compliance with the ADA.

- A state or local agency enforcing a certified code is for practical, but not legal, purposes facilitating compliance with the ADA and helping to eliminate confusion and possible inconsistencies in standards.
- The amount of unnecessary litigation can be reduced, particularly if a state or local code agency has an administrative method of effectively handling complaints concerning violations of its code.

There are tens of thousands of code jurisdictions in the U.S. that enforce some combination of state and local building codes. Some, but not all, of these include accessibility requirements. Although many are based on a model code, there are major variations among the State codes, and among local codes within some States. Design and construction to these codes will not constitute compliance with the ADA, unless the codes impose requirements equal to or more stringent than those ADAAG.

The enforcement of these codes is the responsibility of State or local officials. They usually review building plans and inspect projects at specific intervals during construction to ensure that the construction complies with State and local laws. State and local officials do not have the authority to enforce the ADA on behalf of the Federal government.

Architects, builders, and others involved with design and construction are accustomed to the State and local enforcement system, which lets them know before construction whether they need to make changes to their plans in order to achieve compliance. The ADA relies on the traditional method of case-by-case civil rights enforcement in response to complaints. It does not contemplate Federal ADA inspections similar to those done at the State or local level. The ADA certification provisions will help to moderate the effects of these differences in enforcement procedures and standards.

State or local officials also have no authority to waive ADA requirements. Many State or local codes allow the building official to grant waivers, variances, or other types of exception (e.g., in cases of "undue hardship," "impossibility," or "impracticability"). They may allow compliance by means other than those required by the code if "equivalent facilitation" (see Chapter 2 Section 2.4) is provided. The ADA standards also include some exceptions (e.g., for structural impracticability in new construction) and allow for equivalent facilitation. No individual is authorized under the ADA to grant the exceptions in advance; and the defendant in a lawsuit would have to justify the use of any of those ADA exceptions.

The process to obtain DOJ certification is lengthy and complex. It requires that the code or law submitted for certification to have been formally approved by the issuing body. The local agency must give public notice of its intent to obtain certification and hold onto the record (transcript of the hearing) public hearings.

The document to be submitted must be made available to the public for examination and copying. All material must then be submitted in duplicate to DOJ who will review it and make a preliminary determination of equivalency, after consultation with the ATBCB.

Once a preliminary determination of equivalency is made it will be published in the *Federal Register*, inviting comments for 60 days. An informal hearing may then be held. The comments will be reviewed and the ATBCB will again be consulted and a final determination of equivalency or the certification will be denied. The final action will then be published in the *Federal Register*.

If a jurisdictional authority intends to pursue certification of a local code, they should contact DOJ. The above is only intended to outline the steps for obtaining certification.

A number of elevator code enforcing authorities reported that they have explored certification of their local elevator codes. To date, DOJ has taken the position that they will only entertain applications for certification of a complete accessibility code. The DOJ will also not certify the model codes, but is willing to review them for equivalency with ADAAG. The DOJ rule provides for review of model codes in recognition of the fact that many codes are based on, or incorporate, models or consensus standards developed by nationally recognized organizations. These organizations include, for example, ANSI, BOCA, ICC, SBCCI, and the ICBO.

The model code review process will be informal. DOJ will not necessarily hold a public hearing, but it has the discretion to do so and to ask for public comment. After the review, the DOJ may issue guidance whether and in what respects the model code is consistent with the ADAAG requirement. This guidance will not be binding on any entity or on the DOJ. It will assist in evaluations of individual state or local codes; and it may also serve as a basis for establishing priorities for consideration of individual codes. DOJ is currently reviewing the BOCA and SBCCI model building codes.

As of July 1999, when this edition of *ADA and Building Transportation* was being written, the following was the status of DOJ certified codes.

- On March 29, 1995, the DOJ certified that the State of Washington State Regulations for Barrier-Free Facilities was equivalent to the ADA Standards for Accessible Design.
- On September 23, 1996, the DOJ certified that the Texas Accessibility Standards meet or exceed the requirements of the ADA Standards for Accessible Design.
- On December 12, 1997, DOJ certified that the Maine Human Rights Act, as implemented by the Maine Accessibility Regulations, meets or exceeds the new construction and alterations requirements of Title III of the ADA.
- On May 27, 1998, the DOJ certified the Florida Accessibility Code for Building Construction.

Eight other jurisdictions, including New Mexico, Minnesota, New Jersey, Maryland, California, Indiana, North Carolina, and the County of Hawaii have also submitted codes for certification review. In addition, DOJ is reviewing several model codes. Up-to-date information can be obtained from the DOJ web site http://www.usdoj.gov/crt/ada/certcode.htm

WHAT ACCESSIBILITY REGULATIONS APPLY?

3.

3. WHAT ACCESSIBILITY REGULATIONS APPLY?

3.1 ADA

Title II of ADA applies to public entities. The act defines a public entity as:
(A) any state or local government;
(B) any department, agency, special purpose district, or other instrumentality of a state or states or local government; and
(C) the National Railroad Passenger Corporation, and any commuter authority as defined in Section 103(8) of the Rail Passenger Service Act.

Title III of ADA applies to public accommodations and services operated by private entities. Key terms defined in the act are:

COMMERCE

The term "commerce" means travel, trade, traffic commerce, transportation or communication:
    (A) among the states;
    (B) between any foreign country or any territory or possession and any state; or
    (C) between points in the same state but through another state or foreign country.

COMMERCIAL FACILITIES

The term "commercial facilities" means facilities:
    (A) that are intended for nonresidential use; and
    (B) whose operations will affect commerce.
Such term shall not include facilities that are covered or expressly exempted from coverage under the 1988.

PUBLIC ACCOMMODATION

The following private entities are considered public accommodations for purposes of ADA, if the operations of such entities affect commerce:
    (A) an inn, hotel, motel, or other place of lodging, except for an establishment located within a building that contains not more than five rooms for rent or hire and that is actually occupied by the proprietor of such establishment as the residence of such proprietor;
    (B) a restaurant, bar, or other establishment serving food or drink;

(C) a motion picture house, theater, concert hall, stadium, or other place of exhibition or entertainment;
(D) an auditorium, convention center, lecture hall, or other place of public gathering;
(E) a bakery, grocery store, clothing store, hardware store, shopping center, or other sales or rental establishment;
(F) laundromat, dry-cleaner, bank, barber shop, beauty shop, travel service, shoe repair service, funeral parlor, gas station, office of an accountant or lawyer, pharmacy, insurance office, professional office of a health care provider, hospital, or other service establishment;
(G) a terminal, depot, or other station used for specified public transportation;
(H) a museum, library, gallery, or other place of public display or collection;
(I) a park, zoo, amusement park, or other place of recreation;
(J) a nursery, elementary, secondary, undergraduate, or postgraduate private school, or other place of education;
(K) a day care center, senior citizen center, homeless shelter, food bank, adoption agency, or other social service center establishment; and
(L) a gymnasium, health spa, bowling alley, golf course, or other place of exercise or recreation.

ADA Section 307 exempts private clubs and religious organizations from compliance with Title III. It reads as follows:

"The provisions of this title shall not apply to private clubs or establishments exempted from coverage under Title II of the Civil Rights Act of 1964 (42 U.S.C. 2000-a(e)) or to religious organizations or entities controlled by religious organizations, including places of worship."

Facilities subject to Title III of ADA must comply with ADAAG. Facilities subject to Title II of ADA must comply with ADAAG or UFAS.

See Chapter 2 of this Handbook for an in-depth discussion of where ADA is applicable and not applicable.

3.2 *FAIR HOUSING AMENDMENTS ACT*

Title VIII of the Civil Rights Act of 1968, prohibits discrimination in the sale, rental and financing of dwellings based on race, color, religion, sex or national origin. The FHAA, enacted on September 13, 1988 and effective March 12, 1989, expanded the coverage of Title VIII to prohibit discriminatory housing practices against persons with disabilities and families.

The FHAA applies to most dwellings, residential occupancies, including condominiums and cooperatives. FHAA typically does not apply to a single family house or to a dwelling intended to be occupied by no more than four families, provided the owner maintains and occupies one unit as their residence.

See Chapter 1, Section 1.4 for an indepth review of the FHAA.

3.3 UNIFORM FEDERAL ACCESSIBILITY STANDARDS

The Architectural Barriers Act of 1968 and amendments require all buildings designed, constructed, altered or leased with federal funds to conform to UFAS. Section 504 of the Rehabilitation Act of 1973 stipulated that any program or activity receiving federal financial assistance had to be accessible to everyone. Thus, any institution that received federal funds, even if those funds are not for construction purposes, would be obligated to comply with UFAS.

At the time this second edition is being written, facilities of state and local governments including departments, agencies, special purpose districts, etc. may chose to follow UFAS in lieu of ADAAG.

See Chapter 1, Section 1.3 for indepth review of UFAS.

3.4 MODEL BUILDING CODES

Most building codes in the U.S. are based on one of the three model building codes:
- National Building Code, Published by BOCA
- Standard Building Code, Published by SBCCI
- Uniform Building Code, Published by ICBO

The CABO which in 1998 was merged into the ICC assumed the secretariat of the ANSI A117 Committee in late 1988 or early 1989. A revision to ANSI A117.1 was initiated in July 1989. From the onset, a primary goal of CABO was to produce a standard that would be compatible with the model building codes and to provide enforceable criteria. Throughout the ANSI A117.1 revision process, the Committee endeavored to eliminate application criteria. The application criteria (scoping requirements) tell where, when, and to what extent the ANSI A117.1 requirements would apply.

Concurrently, during the revision of ANSI A117.1, the CABO Board for the Coordination of the Model Codes (BCMC) was preparing model application criteria (scoping requirements) for adoption by the model building codes. The scoping requirements were intended to give the authority having jurisdiction a compatible set of application and technical criteria. The BCMC held numerous public meetings and hearings during the process to receive written and verbal comments regarding proposed recommendations. The last hearing was in June 1992 and a final report was approved for submittal to each of the model building code organizations. During the fall of 1992 the BCMC accessibility proposal was presented to each of the model building code organizations. The three model building code organizations at their respective business meetings accepted the BCMC accessibility proposal. Thus, in the 1993 edition or supplement of the respective model building codes, compliance with CABO/ANSI A117.1-1992 was required. Key scoping provisions, of interest, that appear in the model building codes are as follows:

DEFINITIONS

- ACCESSIBLE MEANS OF EGRESS - a path of travel, usable by a mobility impaired person, that leads to a public way.
- AREA OF REFUGE - an area with direct access to an exit or an elevator where persons unable to use stairs can remain temporarily in safety to await instructions or assistance during emergency evacuation.
- TECHNICALLY INFEASIBLE - an alteration of a building or a facility that has little likelihood of being accomplished because existing structural conditions would require removing or altering a load-bearing member which is an essential part of the structural frame, or because other existing physical or site constraints prohibit modification or addition of elements, spaces, or features which are in full and strict compliance with the minimum requirements for new construction and which are necessary to provide accessibility.

NEW CONSTRUCTION

- All passenger elevators on an accessible route shall be accessible. EXCEPTION: *Elevators within a dwelling unit.*
- Platform (wheelchair) lifts shall not be part of a required accessible route in new construction.
- All required accessible spaces shall be provided with not less than one accessible means of egress. Where more than one means of egress is required from any required accessible space, each accessible portion of the space shall be served by not less than two accessible means of egress.
- Each accessible means of egress shall be continuous from each required accessible occupied area to a public way and shall include accessible routes, ramps, exit stairs, elevators, horizontal exits or smoke barriers.
- An elevator to be considered part of an accessible means of egress shall comply with the requirement of Section 211 of ASME/ANSI

ACCESSIBILITY REGULATIONS BY BUILDING OCCUPANCY

Occupancy	Accessibility Regulations [see notes]			
	ADAAG	FHAA	UFAS	A117.1-1992/1998
Assembly, Theaters with and without Stage	6, 7 & 8		1 & 2	10.00
Assembly, Nightclubs	6 & 7		1.00	10.00
Assembly, Restaurants	6, 7 & 8		1 & 5	10.00
Assembly, Churches	3.00		4.00	10.00
Business	6, 8 & 11		2 & 5	10.00
Educational	6, 8 & 7		1, 2 & 5	10.00
Factory and Industrial	7 & 11		1.00	10.00
High Hazard	7 & 11		1.00	10.00
Institutional	7 & 11		2, 4 & 5	10.00
Mercantile	7 & 11		1.00	10.00
Hotel	7, 8 & 11		1.00	10.00
Residential, Multiple Family	9.00	12.00	1, 4 & 5	10.00
Storage	7 & 11		1, 4 & 5	10.00

Notes:
1. Applicable if federal agencies and/or federal financial assistance provided. (i.e. buildings constructed with HUD Community Development Block Grant Funds). See Chapter 4, Section 4.3.
2. State and local governments may choose to follow UFAS in lieu of ADAAG. See Chapter 4, Section 4.3.
3. Applicable if facility rented to private entity, which operates a place of public accommodation. See Chapter 1, Section 1.5.
4. Applicable if federal financial assistance provided to entity, which rents or uses facility. See Chapter 4, Section 4.3.
5. Applicable if federal, state or local government is owner or tenant. See Chapter 4, Section 4.3.
6. ADAAG not applicable if facility owned and operated by religious entity. See Chapter 1, Section 1.5.
7. ADAAG applicable to facility if operated as a place of public accommodations. See Chapter 1, Section 1.5.
8. ADAAG may not be applicable if facility is owned by state or local government. See Chapter 1, Section 1.5.
9. Facilities with places of public accommodations may be required to comply with ADAAG. See Chapter 1, Section 1.5.
10. Applicable only if adopted by the authority having jurisdiction.
11. ADAAG applicable in commercial facilities. See Chapter 1, Section 1.5.
12. See Chapter 1, Section 1.4

REVIEW OF REGULATIONS

4.

4. REVIEW OF ACCESSIBILITY REGULATIONS

4.1 ANALYSIS OF ELEVATOR REGULATIONS BY SECTION

The following format is used for the Section-by-Section analysis of the regulations. The ADAAG Section is printed, followed by a description of variations in CABO/ANSI A117.1-1992, ANSI A117.1-1986 and UFAS. The **most stringent requirements** for new elevators are **printed in bold** type. ADAAG, ANSI A117.1 and UFAS also include accessibility requirements for existing elevators. Where existing elevator requirements are not the same as those for new elevators, the differences are explained.

ICC/ANSI A117.1-1998 introduced a new numbering scheme (See Chapter 6) for the requirements as well as requirements for LU/LA elevators "destination-oriented elevator system." A destination-oriented elevator system is defined as:

> An elevator system that provides lobby controls to select destination floors, lobby indicators designating which elevator to board, and a car indicator designating the floors at which the car will stop.

Where destination-oriented elevator requirements are not the same as those for new elevators the differences are explained. LU/LA elevators are not included in this analysis as they are not recognized by nor conform to ADAAG.

By adhering to the most stringent requirements, you will be in compliance with the ADAAG, ANSI A117.1-1986, CABO/ANSI A117.1-1992, ICC/ANSI A117.1-1998 and UFAS. Finally, background information and explanatory commentary are presented to assist in understanding the intent of the requirement.

State and local accessibility requirements have not been factored into the analysis. They should be adhered to if more stringent. All the referenced standards have Appendix material, which is considered advisory and non-binding The Appendix material is intended to clarify the position of the organization that wrote the requirements. However, the Appendix material while not mandatory can be used to justify a good faith effort to comply with the respective requirements. The text of the elevator requirements in ADAAG, ANSI A117.1-1986, CABO/ANSI A117.1-1992, ICC/ANSI A117.1-1998 and UFAS is reproduced in a comparison chart in Chapter 6.

4.1.1 GENERAL

<u>ADAAG</u>

4.10.1 General. Accessible elevators shall be on an accessible route and shall comply with 4.10 and with the ASME A17.1-1990,

Safety Code for Elevators and Escalators. Freight elevators shall not be considered as meeting the requirements of this section unless the only elevators provided are used as combination passenger and freight elevators for the public and employees.

ICC/ANSI A117.1-1998

The requirements are essentially identical. The standard requires the elevator to comply with the **Safety Code for Elevators and Escalators ASME/ANSI A117.1-1996** and to be **passenger elevators**. Existing elevators that are made accessible are required to be accessible passenger elevators and comply with the following requirements:

> 407.5.8 Identification. Elevators that comply with Section 407.5 shall be clearly identified with the International Symbol of Accessibility complying with Section 703.7, unless all elevators in the building are accessible.

Existing destination-oriented elevators are required to comply with the requirements for new destination-oriented elevators.

CABO/ANSI A117.1-1992

The requirements are essentially identical with one exception. The Standard requires the elevator to comply with ASME/ANSI A17.1-1990 including Addenda ASME/ANSI A17.1a-1991. Existing passenger elevators that are made accessible shall comply with the following requirements:

> 4.10.2.1 All elevators that are programmed to respond to the same hall call control as the required accessible elevator shall comply with the requirement of 4.10.2.
> 4.10.2.7 Identification. Elevators that comply with this standard shall be clearly identified with the international symbol of accessibility, unless all elevators in the building are accessible. See Fig. 4.28.8.

ANSI A117.1-1986

The requirements are essentially identical. ANSI/ASME A17.1-1984 including Supplement A17.1a-1985 is referenced. A statement is also included on the use of residential elevators and wheelchair lifts, when approved by the authority having jurisdiction. ADAAG and CABO/ANSI A117.1-1992 do not permit the use of private residence elevators. See ADAAG 4.1.3(5) (new construction) and § 4.1.6(3)(g) (alterations), for locations where wheelchair lifts are permitted in lieu of elevators.

UFAS

See explanation under ANSI A117.1-1986. ANSI A17.1-1978 including Supplement A17.1a-1979 is reference.

OVERVIEW

Freight elevators that are used to meet the above requirements must conform to the requirements of ASME A17.1, Rule 207.4, Carrying of Passengers on Freight Elevators. ICC/ANSI A117.1 does not recognize the use of freight elevators even if they comply with ASME A17.1, Rule 207.4. The above regulations typically exempt freight elevators. However, Title I of ADA, which addresses employment, may require a freight elevator to be accessible. The employment provisions in Title I would require that reasonable accommodations be made to a freight elevator to accommodate a freight operator, that has a disability.

4.1.2 AUTOMATIC OPERATION

ADAAG

4.10.2 Automatic Operation. Elevator operation shall be automatic. Each car shall be equipped with a self-leveling feature that will automatically bring the car to floor landings within a tolerance of 1/2 in. (13 mm) under rated loading to zero loading conditions. This self-leveling feature shall be automatic and independent of the operating device and shall correct the overtravel or undertravel.

ICC/ANSI A117.1-1998

The requirements are essentially the same except the self-leveling feature is required to **maintain the car at the floor landing within the 1/2 in. (13 mm) tolerance.** The requirement for existing passenger elevators that are made accessible are the same as new elevators. A releveling feature is required.

CABO/ANSI A117.1-1992

The requirements are identical. Existing passenger elevators that made accessible must comply with the requirement in § 4.10.2.

ANSI A117.1-1986

The requirements are identical. Existing equipment is not addressed.

UFAS

The requirements are identical.

OVERVIEW

Originally, there was a demand from the disability community that the car be perfectly level with the landing. The elevator industry, however, insisted that it required a tolerance of plus or minus 1/2 in. It was pointed out that many regular door thresholds were a half inch, or more, above floor level and could be negotiated by wheelchair users. As a further argument, the interested parties were invited to ride floor-to-floor in elevators that were nearly always available at the building site. The differences in building construction from one floor to another, as well as the leveling tolerance of the elevators, were pointed out. This hands-on approach convinced the interested parties that the plus or minus 1/2 in. was reasonable and would not present an insurmountable obstacle to the disabled.

The ATBCB staff has rendered an informal non-binding opinion (See Chapter 5) that the ADAAG provision does not require releveling. The requirement only addresses leveling when you stop at a landing.

I believe the intent of ANSI A117.1-1980, which ADAAG used as the basis for its regulations, required releveling. Clearly, the intent is that a disabled person should not have to negotiate a mislevel exceeding 1/2 in. Suppose a car with any load levels within the 1/2 in. required. Let's say that the car stops 3/8 in. above the floor with 3/4 load in the car. Once the doors open, a disabled person would only have to negotiate a 3/8 in. mislevel. As passengers exit, the load on the hoist ropes reduces and the corresponding rope stretch reduces. The car now moves up, say another 3/8 in. due to the load reduction. Unless the elevator relevels, the car would be 3/4 in. above the landing sill. If another disabled person approaches the open door, they will have to negotiate a 3/4 in. mislevel. In my opinion, this is contrary to the original intent. I recommend that provisions be provided to maintain the 1/2 in. level after arrival at the floor. ICC/ANSI A117.1-1998 clarifies that the car must be provided with a releveling feature. The plus or minus 1/2 in. must be maintained when the doors are open.

The term "operating device" is defined in the ASME A17.1 Code as: "The car switch, push button, lever, or other manual device used to actuate the control."

Elevators that are attendant operated are not considered accessible as an attendant may not be continuously available.

4.1.3 CALL BUTTONS

ADAAG

4.10.3 Hall Call Buttons. Call buttons in elevator lobbies and halls shall be centered at 42 in. (1065 mm) above the floor. Such call buttons shall have visual signals to indicate when each call is registered and when each call is answered. Call buttons shall be a minimum of 3/4 in. (19 mm) in the smallest dimension. The button designating the up direction shall be on top. (See Fig. 20) Buttons shall be raised or flush. Objects mounted beneath hall call buttons shall not project into the elevator lobby more than 4 in. (100 mm).

ICC/ANSI A117.1-1998

They are essentially identical except **objects mounted beneath hall call buttons shall not protrude more than 1 in. (25 mm)**. The standard specifies the height the buttons can be located above the floor, and requires the space in front of the button to be accessible.

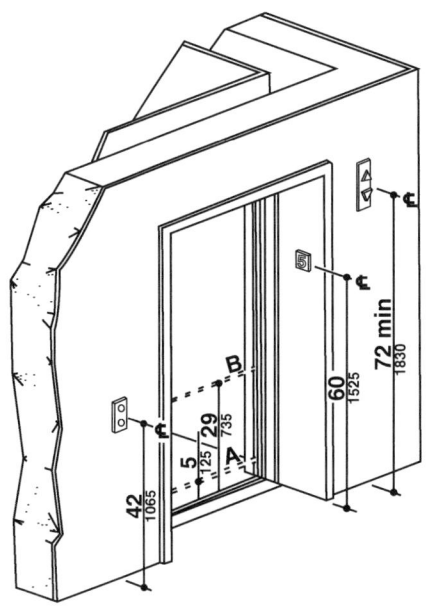

NOTE: The automatic door device is activated if an object passes through either line A or line B. Line A and line B represent the vertical locations of the door reopening device not requiring contact.

**Fig. 20
Hoistway and Elevator Entrances**
Reproduced from ADAAG

> 407.2.2 Call Buttons. Call buttons in elevator lobbies and halls shall be 35 inches (890 mm) minimum and 48 inches (1220 mm) maximum above the floor or ground, measured to the centerline of the buttons. A clear floor or ground space complying with Section 305 shall be provided.

The requirements for existing passenger elevators that are more accessible are identical. In addition, telephone type keypads complying with Section 407.2.2 are permitted.

The requirements for destination-oriented elevators are identical to the requirement for new installations plus the following:

> Destination-oriented elevator systems shall have a keypad or other means for the entry of destination information. Keypads, if provided, shall be in a standard telephone keypad arrangement, and shall be identified by characters complying with Section 703.4. The number five key shall have a single raised dot. The dot shall be 0.118 inch (3 mm) to 0.120 inch (3.05 mm) base diameter, and in other aspects comply with Table 703.5. Destination-oriented

elevator systems shall be provided with visual and audible signals which indicate which elevator car to enter. Characters shall be centered on the corresponding keypad button. A display shall be provided in the car with visible indicators to show registered car destinations. The visible indication shall extinguish when the car arrives at the designated floor. A standard five-pointed star shall be used to indicate the main entry floor.

CABO/ANSI A117.1-1992

The requirements are essentially identical except, there is no requirement for raised or flush buttons. Existing passenger elevators that are made accessible must comply with the following:

> 4.10.2.2 Call Buttons. The top of the hall call buttons shall be located vertically between 35 in. (890 mm) and 54 in. (1370 mm) above the floor when the appropriate floor area specified in 4.2.5 or 4.2.6 is provided. The button that designates the up direction shall be located above the button that designates the down direction.

ANSI A117.1-1986

The requirements are essentially identical, except there is no requirement for raised or flush buttons. Objects mounted beneath hall call buttons are not addressed. Existing equipment is not addressed.

UFAS

The requirements are identical to ADAAG.

OVERVIEW

There are thousands of hall stations throughout the country that comply with the ANSI A117.1 requirements and former NEII handicapped standard that have recessed buttons. The elevator industry has no comments from the users and is unaware of any study that documents that recessed buttons are a problem. Nevertheless, ADAAG, UFAS and ICC/ANSI A117.1-1998 prohibit recessed buttons.

If a button is provided with a ferrule or trim ring, the button surface must either be flush or raised from the adjacent trim ring. A button that is recessed from the adjacent trim ring is considered a recessed button. (See Illustrations in Chapter 5.)

The required 42 in. dimension in all documents except ICC/ANSI A117.1-1998 is a continuation of an old industry standard established by NEII in the 1976 edition of its handicapped standard. The thought at the time was that a button installed in a seven foot door jamb would be at the jamb's center. Thus allowing the side jamb piece of a knock down jamb to be on either the right or left. ICC/ANSI A117.1-1998 recognized the height should be required to be within the accessible reach ranges specified for all other control and operating features.

4.1.4 HALL SIGNALS

ADAAG

> **4.10.4 Hall Lanterns. A visible and audible signal shall be provided at each hoistway entrance to indicate which car is answering a call. Audible signals shall sound once for the up directional and twice for the down direction or shall have verbal annunciators that say "up" or "down." Visible signals shall have the following features:**
> **(1) Hall lantern fixtures shall be mounted so that their centerline is at least 72 in. (1830 mm) above the lobby floor. (See Fig. 20.)**
> **(2) Visual elements shall be at least 21/2 in. (64 mm) in the smallest dimension.**
> **(3) Signals shall be visible from the vicinity of the hall call button (see Fig. 20). In-car lanterns located in cars, visible from the vicinity of hall call buttons, and conforming to the above requirements, shall be acceptable.**

ICC/ANSI A117.1-1998

The requirements are essentially identical plus the following additional requirements for the audible signal.

> Audible signals shall have a frequency of 1,500 Hz maximum. The audible signal or verbal annunciator shall be 10 dBA minimum above ambient, but shall not exceed 80 dBA maximum, measured at the hall call button.

The requirements for existing passenger elevator that are made accessible are as follows:

> 407.5.3 Hall Signals. A visible and audible signal shall be provided at each hoistway entrance to indicate which car is answering a call, except that in-car signals complying with Section 407.2.3 shall be permitted. Audible signals shall sound once for the up direction and twice for the down direction, or shall have verbal annunciators that state the word "up" or "down." If new hall signals are provided, they shall comply with Section 407.2.3.

The requirements for destination-oriented elevators hall signals are as follows:

407.3.2 Hall Signals. A visible and audible signal shall be provided to indicate a car destination corresponding with Section 407.3.1. The audible tone and verbal announcement shall be the same as those given at the call button or call button keypad, if provided. Each elevator in a bank shall have audible and visual means for differentiation.

407.3.2.1. Visible Signals. Visible signals shall comply with Sections 407.3.2.1.1 through 407.3.2.1.3.

407.3.2.1.1 Height. Hall signal fixtures shall be 72 inches (1830 mm) minimum above the floor or ground, measured to the centerline of the fixture.

407.3.2.1.2 Size. The visible signal elements shall be 2 1/2 inches (64 mm) minimum in their smallest dimension.

407.3.2.1.3 Visibility. Signals shall be visible from the floor area adjacent to the hoistway entrance.

CABO/ANSI A117.1-1992

The requirements are essentially identical. Existing passenger elevators that are made accessible must comply with the following:

4.10.2.3 Hall Signals. A visible and audible signal shall be provided at each hoistway entrance to indicate which car is answering a call, except that in-car signals complying with 4.10.1.4 shall be acceptable. Audible signals shall sound once for the up direction and twice for the down direction, or shall have verbal annunciators that state the word "up" or "down." If hall signals are added, they shall comply with 4.10.1.4.

ANSI A117.1-1986

The requirements are essentially identical. Existing equipment is not addressed.

UFAS

The requirements are essentially identical.

OVERVIEW

The height and size of the hall lantern fixture and the audible signal are beneficial to the visually impaired. The requirement "shall be 2 1/2 in. minimum in the smallest dimension" is referring to the minimum height and width of the visual element. The use of in-car lanterns is acceptable, but they will affect elevator door open time. Door open time is calculated from the sounding of the audible signal and illumination of the directional lantern, such that the person awaiting the car is notified which car is responding to their call. If in-car lanterns are used, the door would have to be partially or fully opened before minimum door timing would start. Floor lanterns and audible signals on the other hand can announce the car before it reaches a floor and the door opens.

4.1.5 TACTILE CHARACTERS ON HOISTWAY ENTRANCES

<u>ADAAG</u>

> **4.10.5 Raised and Braille Characters on Hoistway Entrances. All elevator hoistway entrances shall have raised and Braille floor designations provided on both jambs. The centerline of the characters shall be 60 in. (1525 mm) above finish floor. Such characters shall be 2 in. (50 mm) high and shall comply with 4.30.4. Permanently applied plates are acceptable if they are permanently fixed to the jambs. (See Fig. 20).**

<u>ICC/ANSI A117.1-1998</u>

The requirements are essentially identical except **the referenced requirements for signage**. Those reference requirements are discussed in detail in Appendix B. **At the main landing, a tactile star is required in addition to the tactile character.** There is mention of permanently applied plates. As they are not prohibited they are permitted. Existing passenger elevators that are made accessible and destination-oriented elevators are required to comply with the requirement in this §407.2.4. In addition, destination-oriented elevators are required to comply with the following:

> 407.3.5 Elevator Car Identification. In addition to the tactile signs required by Section 407.2.4, a tactile elevator car identification shall be placed immediately below the hoistway entrance floor designation. The characters shall be 2 inches (51 mm) high and shall comply with Section 703.2.

<u>CABO/ANSI A117.1-1992</u>

The requirements are essentially identical, except **the referenced requirements for signage**. There is no mention of permanently applied plates. As they are not prohibited they are permitted. Existing passenger elevators that are made accessible must comply with the requirement in § 4.10.5.

<u>ANSI A117.1-1986</u>

The requirements are essentially identical, except Braille is not required. Existing equipment is not addressed.

UFAS

The requirements are essentially identical, except Braille is not required.

OVERVIEW

In 1985, The Georgia Institute of Technology conducted research that led to the report, "A Multidisciplinary Assessment of the State of the Art of Signage for Blind and Low Vision Persons." With one exception, the criteria established for the car stations was sustained. The exception was on incised (recessed) characters. Georgia Institute found that incised characters were more difficult to identify by the blind for two reasons:

1. The engraving easily fills with debris caused by moisture and dirt on fingertips as well as airborne dust.
2. Many blind people have become blind due to diabetes, which causes a lessening of tactile sensation in the fingertips.

As a result of this study, incised characters are no longer recognized.

The 2 in. dimension provides a character that can be seen by many visually impaired individuals. Characters larger than 2 in. cannot be read tactically. Braille characters can be difficult to read if placed to close to the raised characters. Raised borders can confuse tactile reading of raised characters and Braille.

4.1.6 DOORS

ADAAG

4.10.6* Door Protective and Reopening Device. Elevator doors shall open and close automatically. They shall be provided with a reopening device that will stop and reopen a car door and hoistway door automatically if the door becomes obstructed by an object or person. The device shall be capable of completing these operations without requiring contact for an obstruction passing through the opening of heights of 5 in. and 29 in. (125 mm and 735 mm) above finish floor (see Fig. 20). Door reopening devices shall remain effective for at least 20 sec. After such an interval, doors may close in accordance with the requirements of ASME A17.1-1990.

The asterisk (*) following the Section number indicates that there is non-binding advisory Appendix material. The Appendix material is reproduced in Chapter 6. The following provision for an alteration to an existing elevator is specified in § 4.1.6(3)(c):

> (i) If safety door edges are provided in existing automatic elevators, automatic door reopening devices may be omitted (see 4.10.6).

ICC/ANSI A117.1-1998

The requirements are essentially identical except **doors must be the horizontal type** and you must comply with **ASME/ANSI A17.1-1996** Existing Passenger Elevators that are made accessible must comply with the following:

> 407.5.4 Doors. Doors shall comply with Section 407.5.4.1 or 407.5.4.2.
>
> 407.5.4.1 Power Operated Doors. Power operated horizontally sliding car and hoistway doors opened and closed by automatic means shall comply with Section 407.2.5.
>
> 407.5.4.2 Manually Operated Doors. Existing manually operated hoistway swinging doors shall comply with Sections 404.2.3 and 404.2.9. A power operated car door that opens and maintains a 32 inch (815 mm) minimum clear width shall be provided. Closing of the car door shall not be initiated until the hoistway door is closed. Car gates are prohibited.

Destination-oriented elevators must comply with the requirements for new elevators in §407.2.6.

CABO/ANSI A117.1-1992

The requirements are essentially identical except, you must comply with ASME A17.1-1990 including Addenda ASME A17.1a-1991. Existing passenger elevators that are made accessible must comply with the following:

> 4.10.2.4 Door Operation. Power operated horizontally sliding car and hoistway doors opened and closed by automatic means shall comply with 4.10.1.6. Existing manually operated hoistway swing doors shall comply with 4.13.5 and 4.13.11. A power operated car door that opens and maintains a 32 in. (815 mm) minimum clear width shall be provided. Closing of the car door shall not be initiated until the hoistway door is closed. Car gates are prohibited.
>
> 4.13.5 Clear Width. Doorways shall have a clear opening of 32 in. (815 mm) minimum with door open 90 degrees. Clear opening shall be measured between the face of door and stop. See Fig. B4.13.5. Openings more than 24 in. (610 mm) deep shall comply with 4.2.1 and 4.3.3.

4.13.11* Door-opening Force. Fire doors shall have the minimum opening force allowable by the appropriate administrative authority. The required force for pushing open or pulling open doors other than fire doors shall be as follows:
- interior hinged door: 5.0 lb (22.2 N) maximum
- sliding/folding door: 5.0 lb (22.2 N) maximum

These forces do not apply to the force required to retract latch bolts or disengage other devices that hold the door in a closed position.

ANSI A117.1-1986

The requirements are essentially identical, except ANSI/ASME A17.1-1984 including Supplement A17.1a-1985 is referenced. Existing equipment is not addressed.

UFAS

The requirements are essentially identical, except ANSI A17.1-1978 including Supplement A17.1a-1979 is referenced. The same provision as in ADAAG for alterations to an existing elevator is specified in § 4.1.6(4)(c)(i).

OVERVIEW

The height requirements for the non-contact door reopening devices have been specified such that the device detects a wheelchair foot rest and arm rest. Some mistakenly believed the requirement for a non-contact reopening device would stop the door under all conditions from striking a person or object in the path of the closing door. Owing to the kinetic nature of the motion, reversal of the closing door is not instantaneous. Until the continued movement of the door is arrested, it is possible that limited movement of the door causes it to come in contact with a person or object in its path. This misconception has been addressed in the Appendix material in ANSI A117.1-1986 and CABO/ANSI A117.1-1992, and in the requirement in ICC/ANSI A117.1-1996.

The required 20 sec. requirement does not mean the doors must remain open for 20 sec. It only mandates that the noncontact door reopening device remains active for 20 sec. or until the door has fully closed.

4.1.7 DOOR AND SIGNAL TIMING FOR HALL CALLS

ADAAG

4.10.7* Door and Signal Timing for Hall Calls. The minimum acceptable time from notification that a car is answering a call until the doors of that car start to close shall be calculated from the following equation:
$$T = D/(1.5 \text{ ft/s}) \text{ or } T = D/(445 \text{ mm/s})$$

Where T total time in sec. and D distance (in feet or millimeters) from a point in the lobby or corridor 60 in. (1525 mm) directly in front of the farthest call button controlling that car to the centerline of its hoistway door (see Fig. 21). For cars with in-car lanterns, T begins when the lantern is visible from the vicinity of hall call buttons and an audible signal is sounded. The minimum acceptable notification time shall be 5 sec.

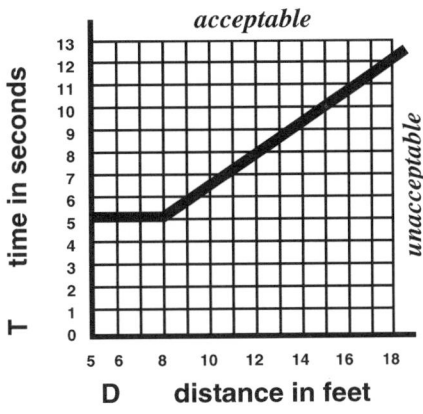

Fig. 21
Graph of Timing Equation
Reproduced from ADAAG

The asterisk (*) following the Section number indicates that there is non-binding advisory Appendix material. The Appendix is reproduced in Chapter 6.

ICC/ANSI A117.1-1998

The requirements are essentially identical. The 5 sec. minimum notification time does not appear in the standards. Existing passenger elevators that are made accessible must comply with the requirements in this §407.2.6. The requirements for destination-oriented elevators are as follows:

> 407.3.6 Door and Signal Timing for Hall Calls. The minimum acceptable time for notification of the car assigned at the keypad until the door starts to close shall be calculated by the following equation, but shall not be less than 5 sec.:
>
> $$T = D/1.5 \text{ ft/s } (D/455 \text{ mm/s})$$
>
> where T = total time in sec. and D = distance in feet (millimeters) from the keypad to the centerline of the assigned hoistway door.

CABO/ANSI A117.1-1992

The requirements are essentially identical. Existing passenger elevators that are made accessible must comply with the requirement in § 4.10.7.

ANSI A117.1-1986

The requirements are essentially identical. Existing equipment is not addressed.

UFAS

The requirements are essentially identical.

OVERVIEW

The requirements do not specify a minimum door open time, but rather a notification time. The requirements give sufficient time to gain access to an elevator. They are based on studies made at Syracuse University in the mid 1970s, which indicated that the disabled needed 1.5 sec for each 1 ft of travel. This requirement gives the disabled time to move into the path of the car door. The provision for door protective and reopening will provide sufficient time to move through the open door.

4.1.8 DOOR DELAY FOR CAR CALLS

ADAAG

4.10.8 Door Delay for Car Calls. The minimum time for elevator doors to remain fully open in response to a car call shall be 3 sec.

ICC/ANSI A117.1-1998

The requirements are essentially identical. Existing passenger elevators that are made accessible and destination-oriented elevators must comply with the requirements for new elevators in §407.2.7.

CABO/ANSI A117.1-1992

The requirements are essentially identical. Existing passenger elevators that are made accessible must comply with the requirement in § 4.10.8.

ANSI A117.1-1986

The requirements are identical. Existing equipment is not addressed.

UFAS

The requirements are identical.

OVERVIEW

This is not a notification time but a minimum door open time. The requirement gives sufficient time to move into the path of the car door. The provisions for the door protective and reopening will provide sufficient time to exit the car with the door remaining open.

4.1.9 INSIDE DIMENSIONS OF ELEVATOR CARS

ADAAG

> **4.10.9 Floor Plan of Elevator Cars. The floor area of elevator cars shall provide space for wheelchair users to enter the car, maneuver within reach of controls, and exit from the car. Acceptable door opening and inside dimensions shall be as shown in Fig. 22. The clearance between the car platform sill and the edge of any hoistway landing shall be no greater than 1¼ in. (32 mm).**
>
> The following provisions for an alteration to an existing elevator are specified in § 4.1.6(3)(c):
>
> (ii) Where existing shaft configuration or technical infeasibility prohibits strict compliance with 4.10.9, the minimum car plan dimensions may be reduced by the minimum amount necessary, but in no case shall the inside car area be smaller than 48 in. by 48 in.
>
> (iii) Equivalent facilitation may be provided with an elevator car of different dimensions when usability can be demonstrated and when all other elements required to be accessible comply with the applicable provisions of 4.10. For example, an elevator of 47 in. by 69 in. (1195 mm by 1755 mm) with a door opening on the narrow dimension, could accommodate the standard wheelchair clearances shown in Fig. 4.

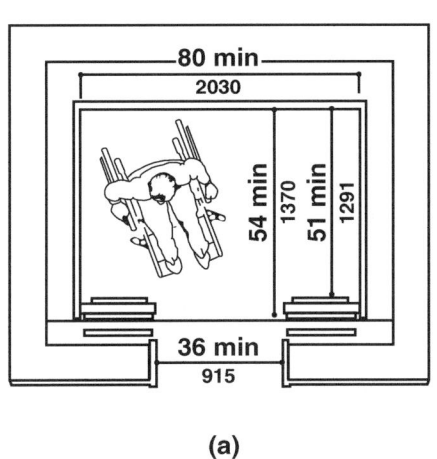
(a)

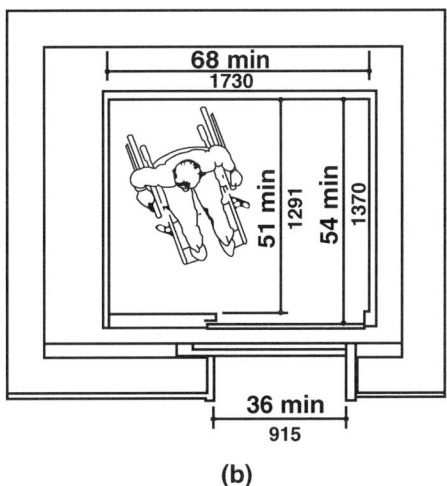
(b)

Fig. 22
Minimum Dimensions of Elevator Cars
Reproduced from ADAAG

ICC/ANSI A117.1-1998

The requirements are essentially identical except they are presented in tabular form as follows:

Table 407.2.8 – Minimum Dimensions of Elevator Cars[1]

Door Location	Door Clear Width	Inside Car, Side to Side	Inside Car, Back Wall to Front Return	Inside Car, Back Wall to Inside Face of Door
Centered	42 inches (1065 mm)	80 inches (2030 mm)	51 inches (1295 mm)	54 inches (1370 mm)
Side (Off Center)	36 inches (915 mm)[2]	68 inches (1725 mm)	51 inches (1295 mm)	54 inches (1370 mm)
Any	36 inches (915 mm)[2]	54 inches (1370 mm)	80 inches (2030 mm)	80 inches (2030 mm)
Any	36 inches (915 mm)[2]	60 inches (1525 mm)	60 inches (1525 mm)	60 inches (1525 mm)

[1] Other car configurations that provide a 36 inch (915 mm) clear door width and a turning space complying with Section 304 with the door closed are permitted.
[2] A tolerance of minus • inch (16 mm) is permitted.

The requirements for existing passenger elevators that are made accessible are as follows:

407.5.5 Inside Dimension of Elevator Cars. The inside dimension of elevator cars shall comply with Section 407.2.8.
EXCEPTION: *Existing car configurations that provide a clear floor area of 16 square feet (1.5 m^2) minimum, and provide 48 inches (1220 mm) minimum inside clear depth and a 26 inch (915 mm) minimum clear width.*

Destination-oriented elevators must comply with the requirements for new elevators in §407.2.8.

CABO/ANSI A117.1-1992

The requirements are essentially identical except that no minimum car size is specified, rather a performance criteria is specified. There also is a non-binding Appendix, which refers the reader to the *NEII Vertical Transportation Standards* (next edition *NEII Building Transportation Standards and Guidelines, NEII-1-2000*) for combination car sizes and door arrangements that provide accessibility. The Appendix is reproduced in Chapter 6.

Existing passenger elevators that are made accessible must comply with the requirement in § 4.10.9

ANSI A117-1986

The requirements are identical. Existing equipment is not addressed.

UFAS

The requirements are identical.

The following provision for alterations to existing buildings is specified in § 4.1.6(4)(c):

(ii) Where existing shaft or structural elements prohibit strict compliance with 4.10.9, then the minimum floor area dimensions may be reduced by the minimum amount necessary, but in no case shall they be less than 48 in. by 48 in. (1220 mm by 1220 mm).

OVERVIEW

The sill-to-sill running clearance received much attention in the early 1970s. Persons using wheelchairs wanted the clearance to be as close to zero as possible. A large gap can be cumbersome to transverse with a wheelchair, especially for the front wheels. It wasn't until the industry explained and demonstrated the need for door operating mechanism (operators, interlocks, etc.) clearances and building construction tolerances (variations) that standard running clearances were agreed upon. In the 1970s, NEII tried to convince the A117.1 Committee to include requirements for one elevator to be of a size to accept a stretcher. The efforts were unsuccessful because the A117.1 standard addresses accessibility to buildings and facilities, and stretchers are not a means of accessibility. The access requirements for stretchers were directly responsible for changing the layout requirements for the 2000-lb elevator. It was found the 2500-lb configuration with 42 in. side opening doors could accommodate a 76 in. by 24 in. ambulance type stretcher in the horizontal position. Since many buildings require two car elevator systems, but not necessarily two 2500-lb cars, the 2000-lb car was reconfigured from a 6 ft 4 in. by 4 ft 5 in. platform to a 6 ft 0 in. by 5 ft 0 in. platform. This provided a better use of building space since it eliminated any wall offset within the hoistway interior. The NEII stretcher requirements did eventually form the basis for inclusion of requirement in the high rise sections of the model building codes.

There was also a strong movement in the early 1970s to eliminate center-opening doors. It was generally felt by the disabled that side-opening doors in conjunction with a side-wall-mounted car station would provide the wheelchair user with direct access to the station as they entered the elevator. The philosophy could not be criticized, but it was pointed out that side-opening doors would result in less than satisfactory handling of a major building's population and, in fact, it could necessitate additional elevators. The ultimate resolution was to continue the use of center-opening doors.

Until ADAAG was published, cars with inside dimensions less than shown in Fig. 22 were considered as being in compliance, provided they met the provisions in § 4.2.3. Section 4.2.3 states that you need a clear space 60 in. in diameter to turn a wheelchair 180 degrees. Thus, elevators with a minimum clear inside dimension of 60 in. by 60 in. were acceptable. The September 6, 1991, *Federal Register* contained the guidelines for transportation facilities. An exception was added for elevators in certain new transit facilities, which permitted cars with a clear floor area in which a 60 in. diameter circle can be inscribed, in lieu of the minimum car dimensions of § 4.10, Fig. 22. In the preamble, the ATBCB responded to a comment that this exception should be in § 4.10 rather than just for elevators in transit facilities. The ATBCB responded, "The Board believes that this exemption should not be permitted under 4.10 (elevators)."

NEII has requested that the ATBCB revisit this issue, as there is no logical reason why the exception provides access in one type of occupancies and not in others. To make an elevator accessible, it is a function of both the inside car dimension and door location, size and operation especially in smaller cars. The space between the door returns is utilized to make a three point return. The *NEII Vertical Transportation Standards* (next edition *NEII Building Transportation Standards and Guidelines, NEII-1-2000*) identifies those car and door arrangements, which comply with the accessibility requirements. Interestingly, the 2500-lb elevator with center opening doors shown in Fig. 22 of ADAAG and UFAS is not accessible. That car arrangement would need a 42 in. door in order for a wheelchair to enter the car, maneuver within reach of controls and exit from the car. There is not enough space with a 36 in. door to make a three point turn with a wheelchair. The ATBCB has been advised of this error by NEII.

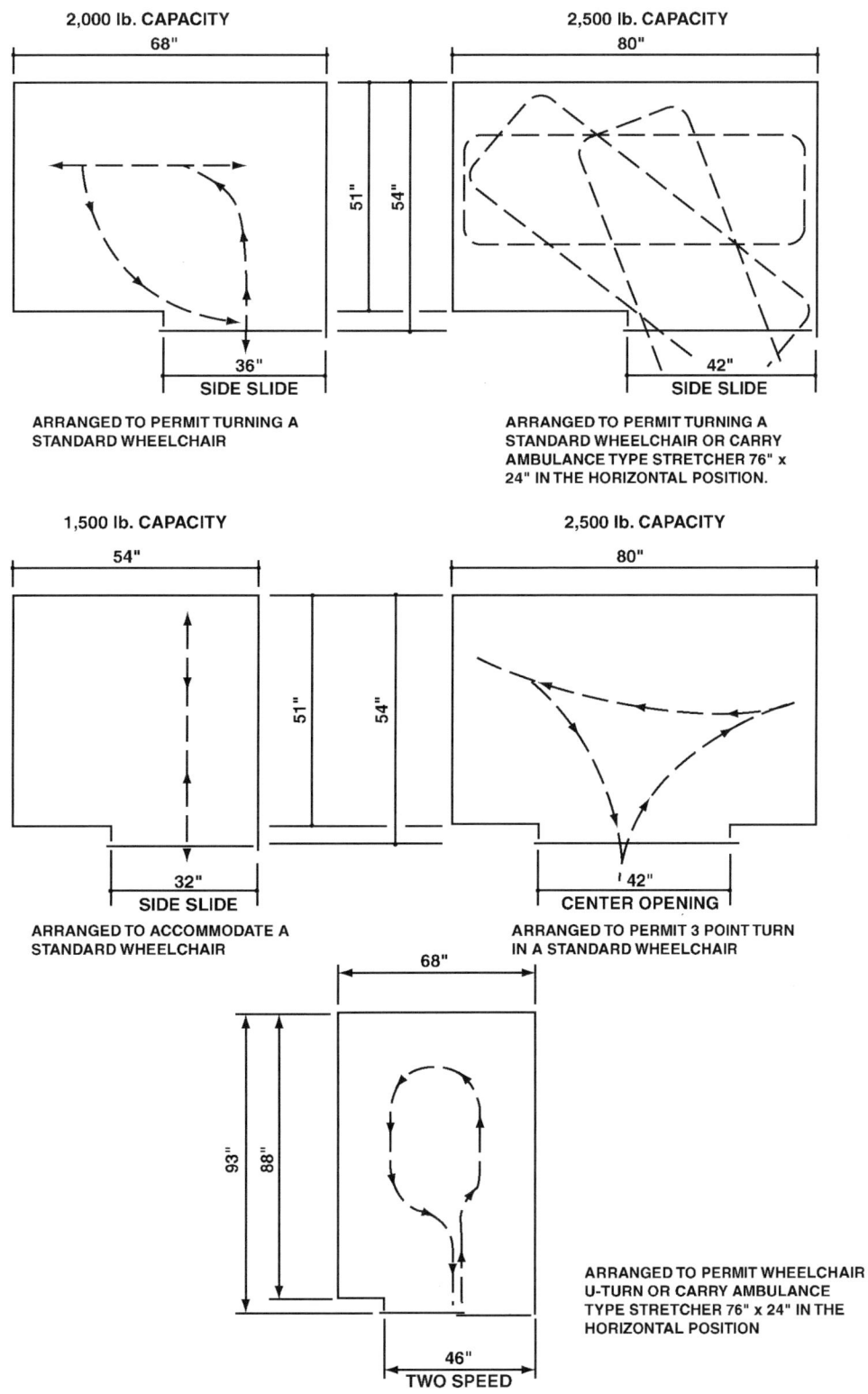

Courtesy National Elevator Industry, Inc., 400 Frank W. Burr Blvd., Teaneck, NJ

4.1.10 FLOOR SURFACES

ADAAG

4.10.10 Floor Surfaces. Floor surfaces shall comply with 4.5.
The applicable requirements in the referenced Section are as follows:

4.5.1* General. Ground and floor surfaces along accessible routes and in accessible rooms and spaces including floors, walks, ramps, stairs, and curb ramps, shall be stable, firm, slip-resistant, and shall comply with 4.5.

4.5.2 Changes in Level. Changes in level up to 1/4 in. (6 mm) may be vertical and without edge treatment (See Fig. 7(c)). Changes in level between 1/4 in. and 1/2 in. (6 mm and 13 mm) shall be beveled with a slope no greater than 1:2 (see Fig. 7(d)). Changes in level greater than 1/2 in. (13 mm) shall be accomplished by means of a ramp that complies with 4.7 or 4.8.

4.5.3* Carpet. If carpet or carpet tile is used on a ground or floor surface, then it shall be securely attached; have a firm cushion, pad or backing, or not cushion or pad; and have a level loop, textured loop, level cut pile, or level cut/uncut pile texture. The maximum pile thickness shall be 1/2 in. (13 mm) (see Fig. 8(f)). Exposed edges of carpet shall be fastened to floor surfaces and have trim along the entire length of the exposed edge. Carpet edge trim shall comply with 4.5.2.

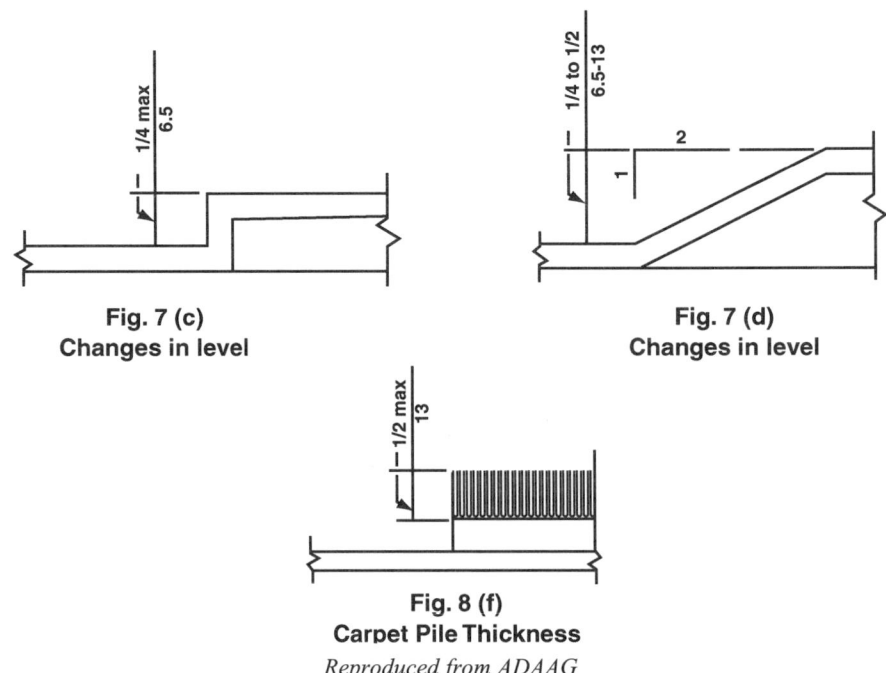

Fig. 7 (c)
Changes in level

Fig. 7 (d)
Changes in level

Fig. 8 (f)
Carpet Pile Thickness

Reproduced from ADAAG

ICC/ANSI A117.1-1998

The requirements are essentially identical. Existing passenger elevators that are made accessible and destination-oriented elevators must comply with the requirements for new elevators in §407.2.9.

CABO/ANSI A117.1-1992

The requirements are essentially identical. Existing passenger elevators that are made accessible must comply with the requirement in § 4.10.10.

ANSI A117.1-1986

The requirements are essentially identical. Existing equipment is not addressed.

UFAS

The requirements are essentially identical.

OVERVIEW

The reference requirements (§ 4.5) require floor surfaces to be stable, firm and slip resistant. Carpets, if provided, must be securely attached and have a firm backing or no backing or pad. Pile height is not to exceed 1/2 in. The exposed edges are to be trimmed.

4.1.11 ILLUMINATION LEVELS

ADAAG

4.10.11 Illumination Levels. The level of illumination at the car controls, platform, and car threshold and landing sill shall be at least 5 footcandles (53.8 lux).

ICC/ANSI A117.1-1998

The requirements are essentially identical. Existing passenger elevators that are made accessible and destination-oriented elevators must comply with the requirements for new elevators in §407.2.10.

CABO/ANSI A117.1-1992

The requirements are identical. Existing passenger elevators that are made accessible must comply with the requirement in § 4.10.11.

ANSI A117.1-1986

The requirements are identical. Existing equipment is not addressed.

UFAS

The requirements are identical.

OVERVIEW

The accessibility requirements are identical to those in ASME A17.1.

4.1.12 CAR CONTROLS

ADAAG

4.10.12* Car Controls. Elevator control panels shall have the following features:
(1) Buttons. All control buttons shall be at least 3/4 in. (19 mm) in their smallest dimension. They shall be raised or flush.
(2) Tactile, Braille, and Visual Control Indicators. All control buttons shall be designated by Braille and by raised standard alphabet characters for letter, Arabic characters for numerals, or standard symbols as shown in Fig. 23(a), and as required in ASME A17.1-1990. Raised and Braille characters and symbols shall comply with 4.30. **The call button for the main entry floor shall be designated by a raised star at the left of the floor designation (see Fig. 23(a)). All raised designations for control buttons shall be placed immediately to the left of the button to which they apply. Applied plates, permanently attached, are an acceptable means to provide raised control designations. Floor buttons shall be provided with visual indicators to show when each call is registered. The visual indicators shall be extinguished when each call is answered.**
(3) Height. All floor buttons shall be no higher than 54 in. (1370 mm) above the finish floor for side approach and 48 in. (1220 mm) for front approach. **Emergency controls, including the emergency alarm and emergency stop, shall be grouped at the bottom of the panel and shall have their centerlines no less than 35 in. (890 mm) above the finish floor (see Figs. 23(a) and (b)).**
(4) Location. Controls shall be located on a front wall if cars have center opening doors, and at the side wall or at the front wall next to the door if cars have side opening doors (see Fig. 23(c) and (d)).

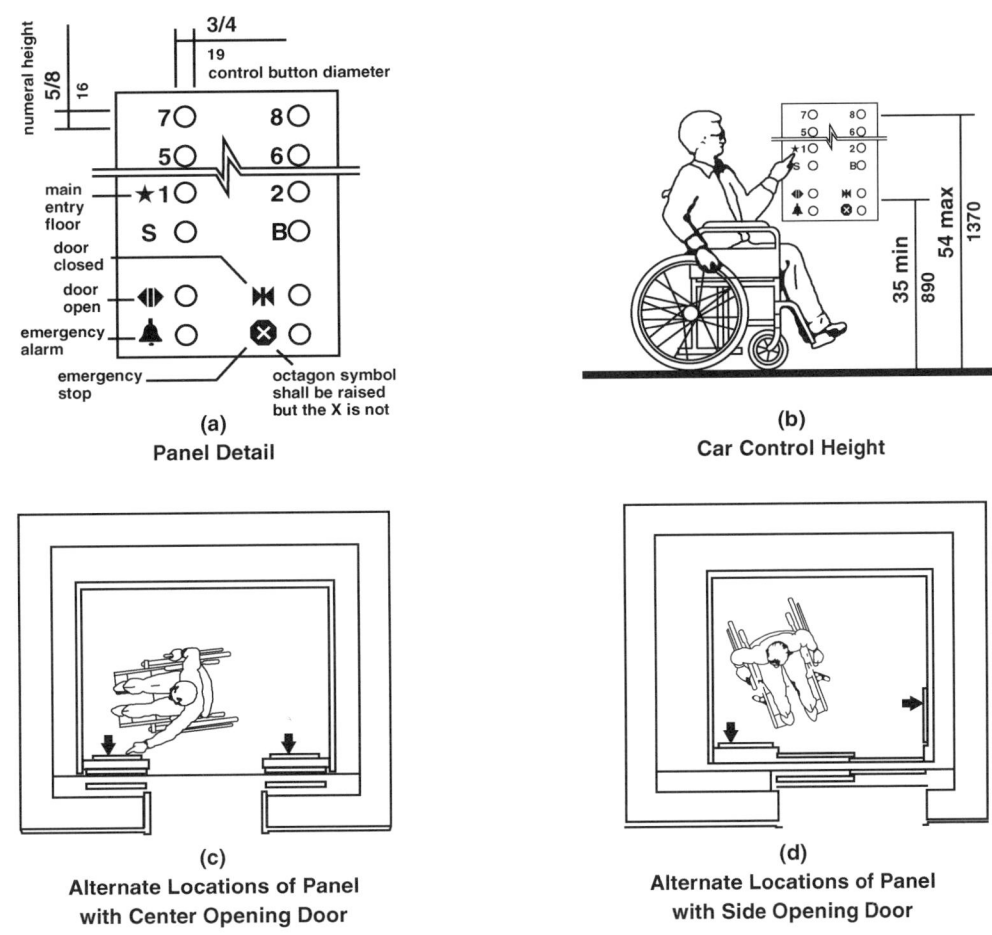

**Fig. 23
Car Controls**
Reproduced from ADAAG

The asterisk (*) following the Section number indicates that there is non-binding advisory Appendix material. The Appendix is reproduced in Chapter 6.

ICC/ANSI A117.1-1998

The requirements are essentially similar. The **reference requirements for signage are more detailed. Braille must be located immediately below the raised control designation. The tactile identification shall be as shown in Appendix B.**

Additional requirements are as follows:

Where two or more columns of buttons are provided they shall read from left to right. Buttons with floor designations shall be 48 inches (1220 mm) maximum above the floor or ground.

EXCEPTION: *Where the elevator serves more than 16 openings and parallel approach is provided, buttons with floor designations shall be 54 inches (1370 mm) maximum above the floor or ground.*

Telephone-style keypads shall be in a standard telephone keypad arrangement, and shall be identified by characters complying with Section 703.4. The number five key shall have a single raised dot. The dot shall be 0.118 inch (3 mm) to 0.120 inch (3.05 mm) base diameter and in other aspects comply with Table 703.5. Characters shall be centered on the corresponding keypad button. A display shall be provided in the car with visible indicators to show registered car destinations. The visible indication shall extinguish when the car arrives at the designated floor. A standard five-pointed star shall be used to indicate the main entry floor.

The requirements for existing passenger elevators that are made accessible are as follows:

407.5.6 Car Controls. Elevator controls shall comply with Sections 407.5.6.1 through 407.5.6.4.

407.5.6.1 Buttons. Car control buttons shall be 3/4 inch (19 mm) minimum in their smallest dimension. Control buttons shall be raised, flush or recessed. Where the car operating panel is changed, control buttons shall comply with Section 407.2.11.1.

407.5.6.2 Designations and Indicators for Control Buttons. All control buttons shall comply with Section 407.2.11.2.
> EXCEPTION: *Where car operating panel construction precludes locating tactile markings to the left of the controls, markings shall be placed as near to the control as possible.*

407.5.6.3 Heights. All buttons with floor designations shall be 54 inches (1370 mm) maximum above the floor for parallel approach and 48 inches (120 mm) maximum above the floor for forward approach. Where the panel is changed, it shall comply with Section 407.2.11.3

407.5.6.4 Operating Panels. Where a new car operating panel complying with the requirements of Section 407.2.11 is provided, existing car operating panels not complying with Section 407.2.11 are not required to be removed.

The requirements for destination-oriented elevators are as follows:

407.3.3 Car Controls. Emergency controls, including the emergency alarm, shall have their centerlines 35 inches (890 mm) minimum and 48 inches (1220 mm) maximum above the floor or ground. Buttons shall be 3/4 inch (19 mm) minimum in their smallest dimension. Buttons shall be raised or flush. Controls shall accommodate a forward reach or side reach complying with Section 308.

CABO/ANSI A117.1-1992

The requirements are essentially identical, however, recessed buttons are permitted. The referenced requirements for signage are more detailed. The Braille must be located immediately below the raised control designation. See signage provisions in Appendix B. There also is an additional requirement:

Control buttons shall be arranged with numbers in ascending order. When two or more columns of buttons are provided they shall read from left to right. See Fig. B4.10.1.12(a).

Existing elevators that are made accessible must comply with the following:

4.10.2.5 Car Controls. Elevator control panels shall have the following features:
    (1) Buttons. Car control buttons shall be 3/4 in. (19 mm) minimum in their smallest dimension. Control buttons shall be raised, flush or recessed.
    (2) Arrangement of Buttons. When the car operating panel is changed, control buttons shall comply with 4.10.1.12(1).
    (3) Tactile and Visual Control Indicators. All control buttons shall comply with 4.10.1.12(2).
EXCEPTION: *When existing car operating panel construction precludes locating tactile markings to the left of the controls, markings shall be placed as near to the control as possible.*
    (4) Height. All floor buttons shall be located 54 in. (1370 mm) maximum above the floor for parallel approach and 48 in. (1220 mm) maximum above the floor for front approach. When the panel is changed, emergency controls, including the emergency alarm, shall comply with 4.10.1.12(3).
    (5) Location. Location of controls shall comply with 4.10.1.12(4).
    (6) Operating Panels. When a new car operating panel conforming to the requirements of 4.10.1.12 is provided, existing car operating panel(s) not conforming to 4.10.1.12 are not required to be removed.

ANSI A117.1-1986

The requirements are essentially identical. Braille identification is not required. A requirement is included that buttons be arranged in ascending order. Existing equipment is not addressed.

UFAS

The requirements are essentially identical except Braille identification is not required.

OVERVIEW

During the research in the early 1970s the area that received the most attention and discussion was the car station and the car call registrations buttons. Extensive testing was conducted at Syracuse University, and additional input was received from wheelchair using members of numerous disability organizations. The optimum minimum and maximum reach capabilities were obtained. This led to the requirements that the emergency buttons in the car station be located 35 in. above the floor and highest floor button 54 in. above the floor. Later studies determined that the maximum height for a forward reach was 48 in. New research by the Little People of America has shown that buttons located above 48 in. may be a problem to people with a short stature. The only exception to the 48 in. maximum height in ICC/ANSI A117.1-1998 is for elevator hall call buttons, and that exception applies only when the elevator serves 16 landings and a parallel approach is provided. The exception was granted due to the limitation of space available for buttons in the front return. It was also determined that a button should be a minimum of 3/4 in. in diameter in order to best serve the needs of the riding public. As an example, some people may have to use the back of the hand to push the button. This dimension was obtained by checking the size of most manufacturers' elevator buttons and finding 3/4 in. was an almost universal industry standard. This size button was then subjected to testing and found to be appropriate. It should be noted that designs encompassed raised, flush or recessed buttons. All met accessibility criteria, none of the designs posing a problem in the registration of calls. However, the Access Board decided that recessed buttons were a problem and prohibited their use in ADAAG and, in the spirit of harmonization, ICC/ANSI A117.1-1998 adopted the same restriction on recessed buttons.

To facilitate call registration by the visually impaired, markings on contrasting background a minimum of 5/8 in. height and raised a minimum of .030 inches were required to be placed adjacent and to the left of the car station buttons. The 5/8 in. height was required since that is the minimum size that can best be discerned tactually by the blind and visually by persons with vision deficiencies, short of legal blindness.

The specifications for side and forward reach are given in § 4.2.5 and § 4.26 of the referenced documents, except ICC/ANSI A117.1-1998 it is in §305.7.2 and §305.7.2. Appendix Fig. A3(a) in ADAAG clarifies the requirements.

To following figures will show that a side reach is not available in a 2000 lb capacity elevator, with the car control station located in the front return.

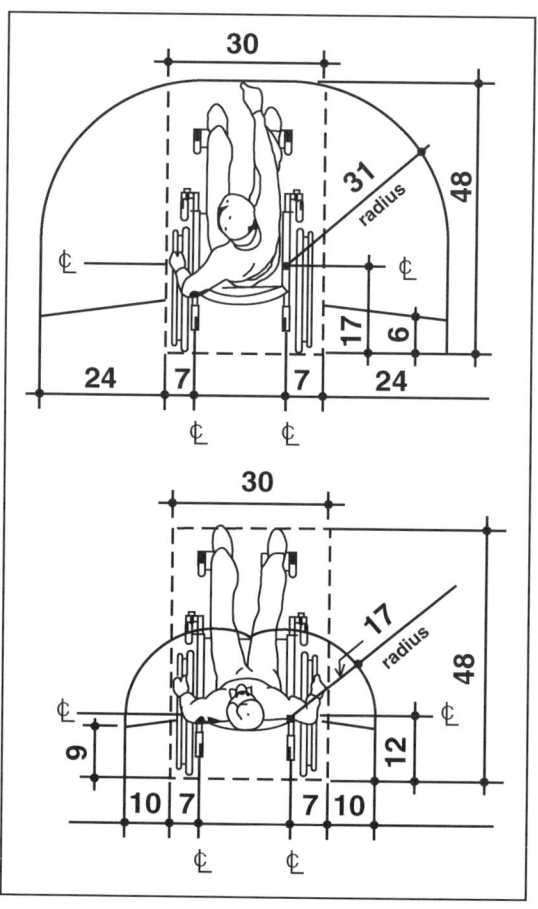

**Fig. A3
Dimensions of Adult-Sized Wheelchairs**
Reproduced from ADAAG

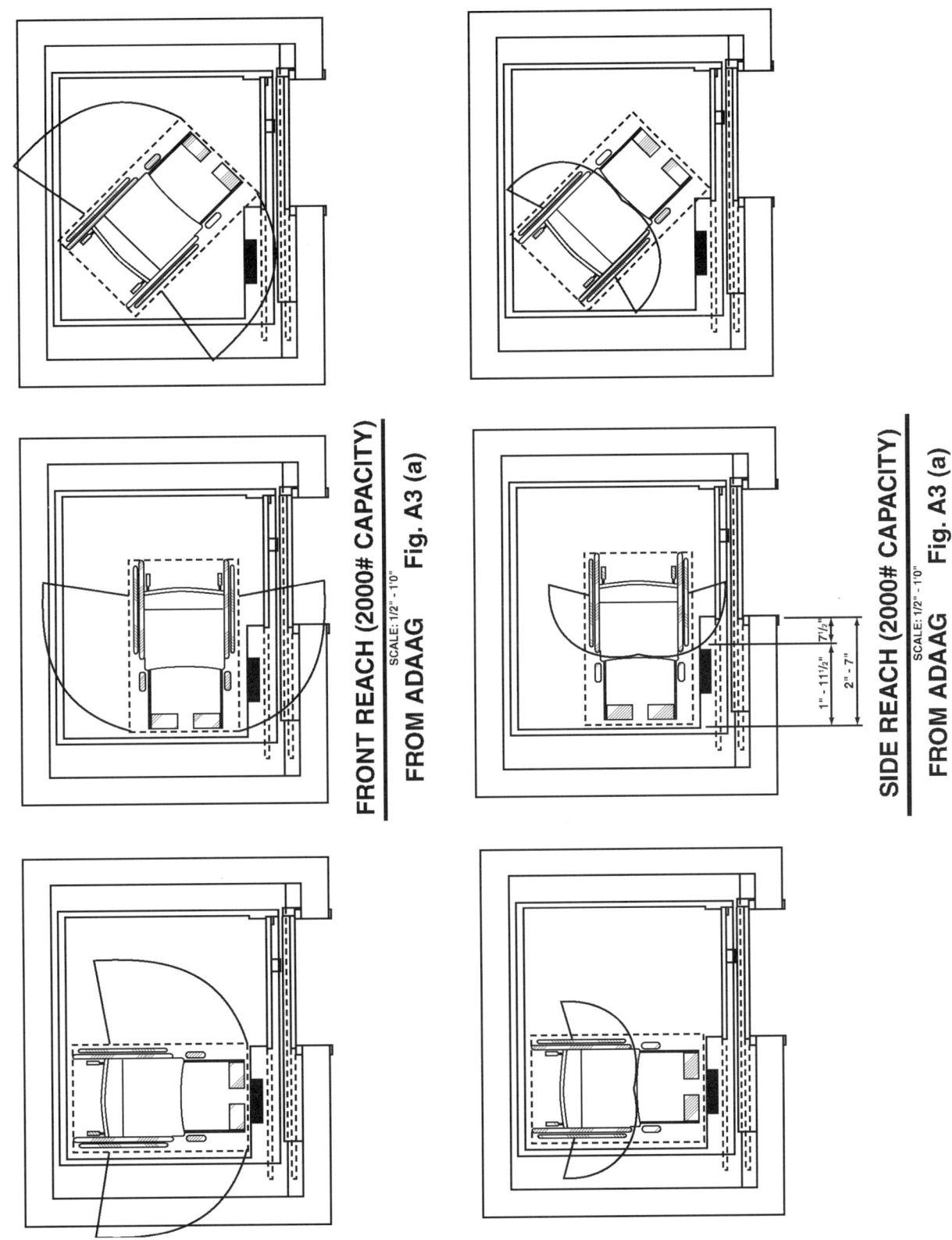

If a button is provided with a ferrule or trim ring, the button surface must either be flush or raised from the adjacent trim ring. A button that is recessed from the adjacent trim ring is considered a recessed button.

The use of a telephone type keypad format is encouraged in the Appendixes of the referenced accessibility standards. ICC/ANSI A117.1-1998 has specific requirements for telephone type keypads. Systems using telephone keypads also incorporate familiar format making use convenient for those with visual impairments. Thus Braille is not required to make recognition more convenient in buildings with basement floors, it will often be desirable to change the "#" symbol to a "..". The dot on the "5" key helps the blind orientate their fingers. The telephone type keypad can be found inside the car or in the hall for destination orientated elevators.

4.1.13 CAR POSITION INDICATORS

<u>ADAAG</u>

> **4.10.13* Car Position Indicators. In elevator cars, a visual car position indicator shall be provided above the car control panel or over the door to show the position of the elevator in the hoistway. As the car passes or stops at a floor served by the elevators, the corresponding numerals shall illuminate,** and an audible signal shall sound. **Numerals shall be a minimum of 1/2 in. (13 mm) high.** The audible signal shall be no less than 20 decibels with a frequency no higher than 1,500 Hz. An automatic verbal announcement of the floor number at which a car stops or which a car passes may be substituted for the audible signal.

The asterisk (*) following the Section number indicates that there is non-binding Appendix material. The Appendix is reproduced in Chapter 6.

<u>ICC/ANSI A117.1-1998</u>

The requirements are essentially identical except for the following additional requirements that apply to the audible signal:

> 407.2.12.2 Audible Indicators. The audible signal shall be 10 dBA minimum above ambient, but shall not exceed 80 dBA maximum, measured at the annunciator. The signal shall be an automatic verbal announcement, which announces the floor at which the car has stopped.
> > EXCEPTION: *For elevators that have a rated speed of 200 fpm (1 m/s) or less, an audible signal with a frequency of 1,500 Hz maximum which sounds as the car passes or stops at a floor served by the elevator shall be permitted.*

The requirements for existing installations that are made accessible are as follows:

> 407.5.7 Car Position Indicators. Where a new car position indicator is provided, the indicator shall comply with Section 407.2.12.

The requirements for destination-oriented elevators are as follows:

> 407.3.4 Car Position Indicators. In elevator cars, audible and visible car location indicators shall be provided.
>
> 407.3.4.1 Visible Indicators. Indicators shall be above the car control panel or above the door. Numerals shall be 1/2 inch (13 mm) high minimum. The visible indicators shall extinguish when the car arrives at the designated floor.
>
> 407.3.4.2 Audible Indicators. **An automatic verbal announcement which announces the floor at which the car has stopped shall be provided. The announcement shall be 10 dBA minimum above ambient and 80 dBA maximum, measured at the annunciator.**
> EXCEPTION: *For elevators that have a rated speed of 200 fpm (1m/s) or less, an audible signal with a frequency of 1,500 Hz maximum which sounds as the car passes or stops at a floor served by the elevator shall be permitted.*

CABO/ANSI A117.1-1992

The requirements are essentially identical. Existing passenger elevators that are made accessible must comply with the following:

> 4.10.2.6 Car Position Indicators. When a new car position indicator is installed, the indicator shall comply with 4.10.1.13.

ANSI A117.1-1986

The requirements are essentially identical. Existing equipment is not addressed.

UFAS

The requirements are essentially identical.

OVERVIEW

The audible signal has been found to be of little value to blind persons on high-speed elevators. They are unable to count the audible signal sounded as each floor is passed. ICC/ANSI A117.1-1998 thus requires a verbal announcement on elevators with a car speed 200 fpm or more. Audible signals are also not effective when floor are not sequentially numbered such as where there is a mezzanine between the 1st and 2nd floors or where there is no 13th floor.

The audible signal is intended to provide a means for the blind to count floors as the elevator moves. The Appendix for the referenced standards except ICC/ANIS A117.1-1998 indicates that a special button may be provided that would activate the audible signal, within the given elevator only for the desired trip, rather than maintaining the audible signal in constant operation.

4.1.14 EMERGENCY COMMUNICATIONS

ADAAG

4.10.14* Emergency Communications. If provided, emergency two-way communications systems between the elevator and a point outside the hoistway shall comply with ASME A17.1-1990. The highest operable part of a two-way communication system shall be a maximum of 48 in. (1220 mm) from the floor of the car. It shall be identified by a raised symbol and lettering complying with 4.30 and located adjacent to the device. If the system uses a handset then the length of the cord from the panel to the handset shall be at least 29 in. (735 mm). If the system is located in a closed compartment the compartment door hardware shall conform to 4.27, Controls and Operating Mechanisms. The emergency intercommunication system shall not require voice communication.

The asterisk (*) following the Section number indicates that there is non-binding Appendix material. The Appendix is reproduced in Chapter 6.

ICC/ANSI A117.1-1998

The requirements are essentially identical except the reference is to **ASME/ANSI A17.1-1996**. An additional requirement is as follows:

If instructions for use are provided, essential information shall be presented in both tactile and visual form complying with Section 703.

Existing elevators that are made accessible and destination-oriented elevators are required to comply with the requirements in this §407.2.13.

CABO/ANSI A117.1-1992

The requirements are essentially identical except the reference is to ASME A17.1-1990 including Addenda A17.1a-1991. Also the communications system can be 54 in.

above the floor if a parallel approach is possible. An additional requirement is also provided:

> If instructions for use are provided, essential information shall be presented in both tactile and visual form.

Existing passenger elevators that are required to be made accessible must comply with the requirement in § 4.10.10.

ANSI A117.1-1986

The requirements are essential identical, except reference is made to ANSI/ASME A17.1-1984 and A17.1a-1985. The height of the communication system and requirements for instructions are the same as CABO/ANSI A117.1-1992. Existing equipment is not addressed.

UFAS

The requirements are essentially identical except reference is made to ANSI A17.1-1978 and A17.1a-1979.

OVERVIEW

The section starts out by stating "if emergency communication is provided," but § 4.10.1 requires adherence to ASME A17.1, thus you have no choice but to provide emergency communications. The Appendix (see Chapter 6) clarifies the requirement. While technically the Appendix material is not mandatory, it must be kept in mind that interpretations of ADAAG will ultimately be made by the Federal Judiciary. Typically, the courts will review the "record" to determine what was intended before rendering a decision. It is the author's opinion that the Appendix will ultimately become the basis for any interpretation by the courts.

The use of a light, indicating the call has been received and responded to, will provide a system that meets the intent of the requirements for the hearing-impaired. The light on the handsfree phone panel should only be illuminated or blink when activated by the recipient of the call. Those who are blind will be able to use the phone to hear that help is on the way. The instructions on the use of the light(s), need not be raised or in Braille as this feature is not being provided for those with a vision impairment.

If a button is provided to initiate the call, § 4.10.12 will require it to be at least 3/4 in. in the smallest direction. Finally, although an illuminated alarm button is now required by ASME A17.1a-1991, it does not by itself meet the full intent of ADAAG.

4.1.15 TRANSPORTATION FACILITIES

ADAAG

10.1 General. Every station, bus stop, bus stop pad, terminal, building or other transportation facility, shall comply with the applicable provisions of 4.1 through 4.35, sections 5 through 9, and the applicable provisions of this section. The exception for elevators in 4.1.3(5), exception 1 and 4.1.6(1)(k) do not apply to a terminal, depot, or other station used for specified public transportation, or an airport passenger terminal, or facilities subject to Title II.
10.3 Fixed Facilities and Stations.
10.3.1 New Construction. New stations in rapid rail, light rail, commuter rail, intercity bus, intercity rail, high speed rail, and other fixed guideway systems (e.g. automated guideway transit, monorails, etc.) shall comply with the following provisions, as applicable:
(17) Where provided, elevators shall be glazed or have transparent panels to allow an unobstructed view both in to and out of the car. Elevators shall comply with 4.10.
EXCEPTION: *Elevator cars with a clear floor area in which a 60 in. diameter circle can be inscribed may be substituted for the minimum car dimensions of 4.10, Fig. 22.*

ICC/ANSI A117.1-1998

There are no similar provisions.

CABO/ANSI A117.1-1992

There are no similar provisions.

ANSI A117.1-1986

There are no similar provisions.

UFAS

There are no similar provision.

OVERVIEW

If the elevator is installed in a fire-rated hoistway, the only way to meet the requirement for an unobstructed view both into and out of the car is with a vision panel. The vision panel must comply with ASME A17.1, Rules 110.7a and 204.2e.

4.2 ANALYSIS OF WHEELCHAIR REGULATIONS BY SECTION

The following format is used for the Section by Section analysis of the regulations. The ADAAG Section is printed, followed by a description of variations in ICC/ANSI A117.1-1998, CABO/ANSI A117.1-1992, ANSI A117.1-1986 and UFAS. The **most stringent requirement** for wheelchair lifts is **printed in bold** type. By adhering to the most stringent requirements, you will be in compliance with the ADAAG, ICC/ANSI A117.1-1998, CABO/ANSI A117.1-1992, ANSI A117.1-1986 and UFAS. Finally, background information and explanatory commentary will be presented to assist in understanding the intent of the requirement.

State and local accessibility requirements have not been factored into the analysis. They should be adhered to if more stringent. All the referenced standards have Appendix material, which is considered advisory and non-binding. The Appendix material is intended to clarify the position of the organization that wrote the requirements. However, the Appendix material while not mandatory, can be used to justify a good faith effort to comply with the respective requirements.

The text of the wheelchair lift requirements in ADAAG, ANSI A117.1-1986, CABO/ANSI A117.1-1992 and UFAS are reproduced in a comparison chart in Chapter 6.

4.2.1 LOCATION

ADAAG

> **4.11.1 Location. Platform lifts (wheelchair lifts) permitted by 4.1 shall comply with the requirement of 4.11.**

ADAAG restricts the use of wheelchair lifts in new construction. They may be used in new construction in lieu of an elevator only under the following conditions:

> **(a) To provide an accessible route to a performing area in an assembly occupancy.**
> **(b) To comply with the wheelchair viewing position line-of-sight and dispersion requirements of 4.33.3.**
> **(c) To provide access to incidental occupiable spaces and rooms which are not open to the general public and which house no more than five persons, including but not limited to equipment control rooms and projection booths.**
> **(d) To provide access where existing site constraints or other constraints make use of a ramp or an elevator infeasible.**

ICC/ANSI A117.1-1998

No similar requirement. See model building codes for scooping requirements. See Chapter 3, Section 3.6.

CABO/ANSI A117.1-1992

No similar requirement. See model building codes for scoping requirement.

ANSI A117.1-1986

This standard does not limit the use of wheelchair (platform) lifts.

UFAS

This standard does not limit the use of wheelchair (platform) lifts.

FHAA

The Fair Housing Accessibility Guidelines do not prohibit the use of wheelchair lifts.

OVERVIEW

ADAAG permits the use of wheelchair lifts in lieu of elevators in existing buildings. The model building codes reference CABO/ANSI A117.1-1992. The 1993 and later edition or supplement of the model building code (NBC, SBC and UBC) does not recognize or permit the use of wheelchair lifts as part of an accessible route in new construction. They are recognized as part of an accessible route in existing construction. See Chapter 3, Section 3.4.

The restricted use of wheelchair lifts may be due to the ASME A17.1 provision that they are for use of people in wheelchairs or seated in a permanently installed seat. ASME A17.1 Inquiry 88-22 addressed this issue as follows:

> Inquiry: 88-22
> Subject: Part XX
> Vertical and Inclined Wheelchair Lifts
> Edition: A17.1-1984
> Question: (4) Re: Part XX Inclined and Vertical Wheelchair Lifts Are the above mentioned devices only to be used by persons using wheelchairs? Does the fact that the word "wheelchair" is stated in the aforementioned titles of this particular national standard preclude the use of these devices by persons with other disabilities?

	This would include individuals who use canes, crutches, walkers, etc. and have limited physical mobility or strength.
Answer:	(4) Inclined wheelchair lifts provided with seats as covered by Rule 2001.6d are intended to carry physically handicapped persons in wheelchairs or in seats. Other inclined wheelchair lifts and vertical wheelchair lifts are intended to carry persons in wheelchairs.

4.2.2 REQUIREMENTS

ADAAG

>**4.11.2* Other Requirements. If platform lifts (wheelchair lifts) are used, they shall comply with 4.2.4, 4.5, 4.27, and ASME A17.1 Safety Code for Elevators and Escalators, Section XX,** 1990.

The asterisk (*) following the Section number indicates that there is non-binding advisory Appendix material. The Appendix is reproduced in Chapter 6.

The applicable requirements in the referenced Sections are as follows:

>**4.2.4* Clear Floor or Ground Space for Wheelchairs.**
>**4.2.4.1 Size and Approach. The minimum clear floor or ground space require to accommodate a single, stationary wheelchair and occupant is 30 in. by 48 in. (760 mm by 1220 mm) (see Fig. 4(a)). The minimum clear floor or ground space for wheelchairs may be positioned for forward or parallel approach to an object (see Fig. 4(b) and (c)). Clear floor or ground space for wheelchairs may be part of the knee space required under some objects.**
>**4.2.4.2 Relationship of Maneuvering Clearance to Wheelchair Spaces. One full unobstructed side of the clear floor or ground space for a wheelchair shall adjoin or overlap an accessible route or adjoin another wheelchair clear floor space. If a clear floor space is located in an alcove or otherwise confined on all or part of three sides, additional maneuvering clearances shall be provided as shown in Fig. 4(d) and (e).**
>**4.2.4.3 Surfaces for Wheelchair Spaces. Clear floor or ground spaces for wheelchairs shall comply with 4.5.**
>**4.5 Ground and Floor Surfaces.**
>**4.5.1* General. Ground and floor surfaces along accessible routes and in accessible rooms and spaces including floors, walks,**

ramps, stairs, and curb ramps, shall be stable, firm, slip-resistant, and shall comply with 4.5.

4.5.2 Changes in Level. Changes in level up to 1/4 in. and 1/2 in. (6 mm and 13 mm) shall be beveled with a slope no greater than 1:2 (see Fig. 7(d)). Changes in level greater than 1/2 in. (13 mm) shall be accomplished by means of a ramp that complies with 4.7 and 4.8.

4.5.3* Carpet. If carpet or carpet tile is used on a ground or floor surface, then it shall be securely attached; have a firm cushion, pad, or backing, or no cushion or pad; and have a level loop, textured loop, level cut pile, or level cut/uncut pile texture. The maximum pile thickness shall be 1/2 in. (13 mm) (see Fig. 8(f)). Exposed edges of carpet shall be fastened to floor surfaces and have trim along the entire length of the exposed edge. Carpet edge trim shall comply with 4.5.2.

4.5.4 Gratings. If gratings are located in walking surfaces, then they shall have spaces no greater than 1/2 in. (13 mm) wide in one direction (see Fig. 8(g)). If gratings have elongated openings, then they shall be placed so that the long dimension is perpendicular to the dominant direction of travel (see Fig. 8(h)).

4.27 Controls and Operating Mechanisms.

4.27.1 General. Controls and operating mechanisms required to be accessible by 4.1 shall comply with 4.27.

4.27.2 Clear Floor Space. Clear floor space complying with 4.2.4 that allows a forward or a parallel approach by a person using a wheelchair shall be provided at controls, dispensers, receptacles, and other operable equipment.

4.27.3* Height. The highest operable part of controls, dispensers, receptacles, and other operable equipment shall be placed within at least one of the reach ranges specified in 4.2.5 and 4.2.6. Electrical and communications system receptacles on walls shall be mounted no less than 15 in. (380 mm) above the floor.

EXCEPTION: *These requirements do not apply where the use of special equipment dictates otherwise or where electrical and communications systems receptacles are not normally intended for use by building occupants.*

4.27.4 Operation. Controls and operating mechanisms shall be operable with one hand and shall not require tight grasping, pinching, or twisting of the wrist. The force required to activate controls shall be no greater than 5 lbf (22.2 N).

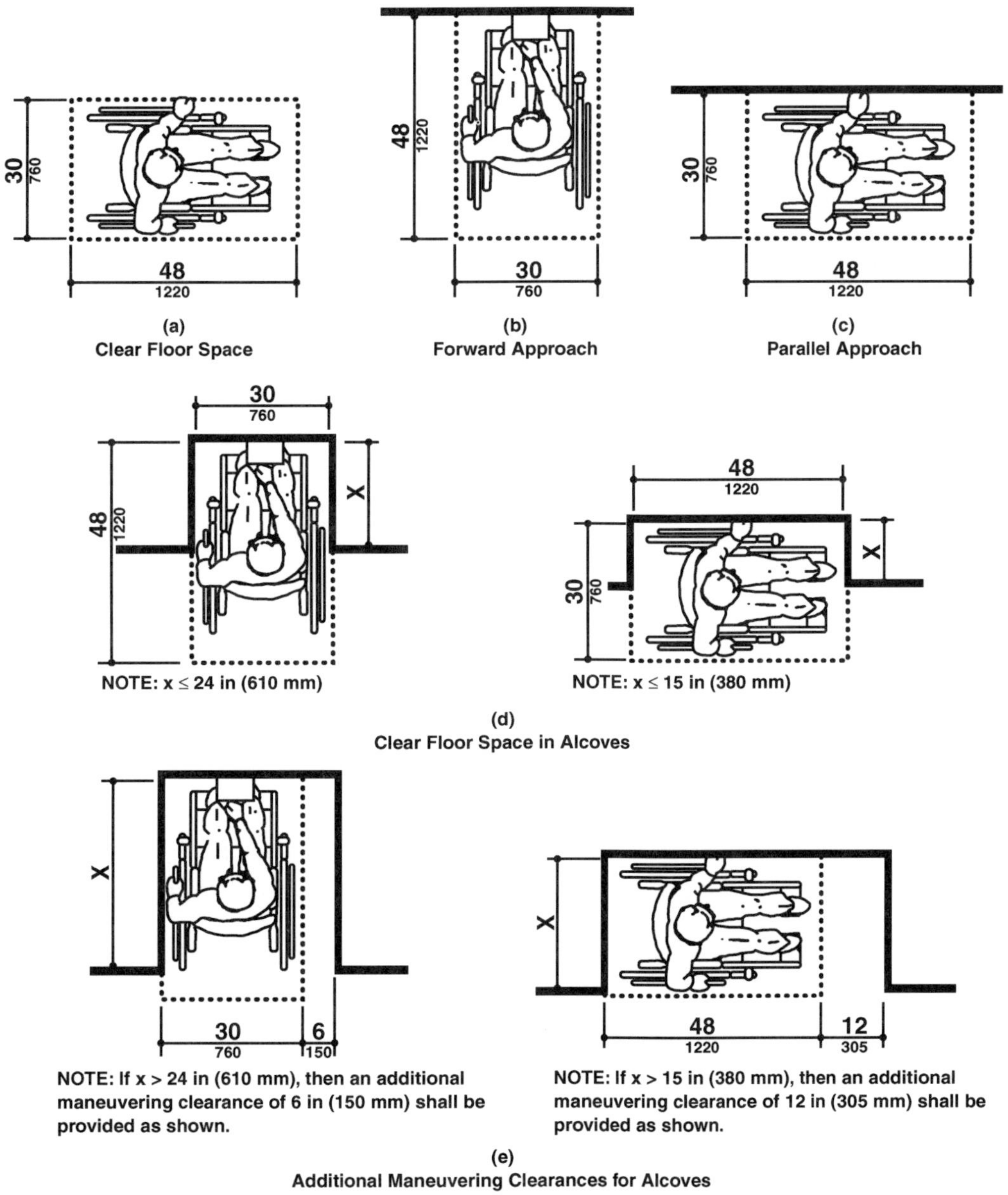

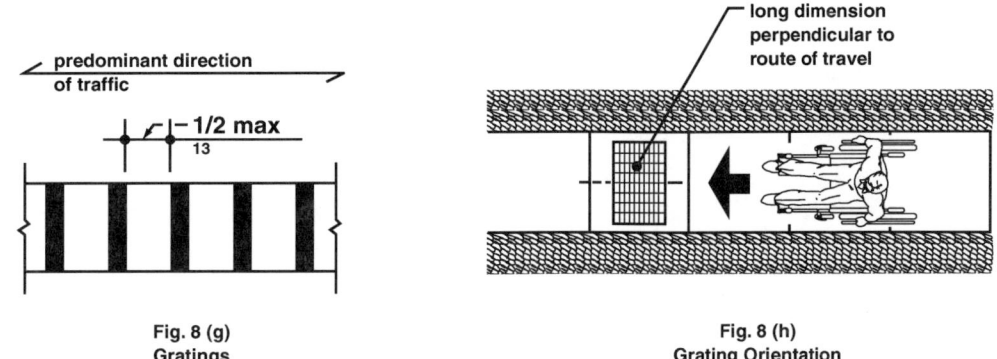

30
760
48
1220
(a)
Clear Floor Space
48
1220
30
760
(b)
Forward Approach
30
760
48
1220
(c)
Parallel Approach
30
760
48
1220
X
NOTE: x ≤ 24 in (610 mm)
48
1220
30
760
X
NOTE: x ≤ 15 in (380 mm)
(d)
Clear Floor Space in Alcoves
X
30
760
6
150
NOTE: If x > 24 in (610 mm), then an additional maneuvering clearance of 6 in (150 mm) shall be provided as shown.
X
48
1220
12
305
NOTE: If x > 15 in (380 mm), then an additional maneuvering clearance of 12 in (305 mm) shall be provided as shown.
(e)
Additional Maneuvering Clearances for Alcoves

Fig. 4
Minimum Clear Floor Space for Wheelchairs

predominant direction
of traffic
1/2 max
13

Fig. 8 (g)
Gratings

long dimension
perpendicular to
route of travel

Fig. 8 (h)
Grating Orientation

Reproduced from ADAAG

ICC/ANSI A117.1-1998

The requirements are essentially identical, except **ASME/ANSI A17.1-1996** is referenced.

CABO/ANSI A117.1-1992

The requirements are essentially identical, except ASME A17.1-1990 including A17.1a-1991 is referenced.

ANSI A117.1-1986

The requirements are essentially identical, except ANSI/ASME A17.1-1984 and A17.1a-1985 is referenced.

UFAS

The requirements are essentially identical, except ASME A17.1 is not referenced. The applicable safety regulations of administrative authorities having jurisdiction are referenced.

OVERVIEW

To comprehend this provision one must review the referenced Sections. ADAAG Section 4.2.4 requires that a wheelchair lift platform be a minimum of 30 in. by 48 in. wherever a clear floor or ground space is specified. The surface area can be on the floor of a building or on the ground as stated but this can also be read as the platform of a lift or elevator. The clear space is also needed on the floor or ground adjacent to the lift to provide maneuvering space for the wheelchair so the user can access the lift.

A 30 in. by 48 in. surface area would accommodate a wheelchair if an entrance to the platform of the lift is from the 30 in. dimension, i.e., a forward approach through a 30 in. opening onto the platform. Inasmuch as the platform would be longer than 24 in., the argument could be put forward that the width should be 36 in. minimum as required by 4.2.1.

If the entrance to a platform lift is to be from the side, i.e., the 48 in. dimension, than additional length would have to be provided in order for the wheelchair user to "park" the wheelchair on the platform. This being the same as maneuvering into an alcove (4.2.4.2).

Smaller platforms restrict the use of the wheelchair lift. In many cases they would not accommodate powered wheelchairs and powered scooters. Other requirements in this Section address forward and side reach [i.e., controls, and usable hardware (doors, etc.)] and maneuvering clearance to access and egress lift. ADAAG Section 4.5 address the requirements for the wheelchair lift platform surface including any ramp used to gain access to the wheelchair lift. This would include the slope of any retractable ramp that is provided on a wheelchair lift. The maximum slope of a ramp in new construction is 1:12. For existing buildings, a slope steeper than 1:18 is not allowed. For slopes steeper than 1:10 but not steeper than 1:8, the maximum rise shall be 3 in. For slopes steeper than 1:12 but not steeper than 1:10, the maximum rise shall be 6 in.

The final ADAAG reference is to Section 4.27 control and operating mechanisms. Control and operating mechanisms must be mounted within reach, which is determined by whether forward or side reach space is provided. See Chapter 4, Section 4.1.12 for clarification of forward and side reach requirements. Quite often the requirement in Section 4.27.4 Operation has been interpreted as not permitting key operation. That interpretation is incorrect. There are no provisions in ADAAG or any of the other standards being reviewed that prohibit keys. If keys were prohibited, a locked door would not be allowed. Security is not addressed by any of the accessibility standards.

The minimum platform size of 30 in. by 48 in. is also specified in ICC/ANSI A117.1-1998, CABO/ANSI A117.1-1992, ANSI A117.1-1986 and UFAS. The other provision reviewed also are contained in CABO/ANSI A117.1-1992, ANSI A117.1-1986 and UFAS, though certain requirements may be more restrictive in one standards versus the other.

The final provision in ADAAG, ICC/ANSI A117.1-1998, CABO/ANSI A117.1-1992 and ANSI A117.1-1986 is that wheelchair lifts must comply with ASME A17.1 Part XX.

4.2.3 ENTRANCE AND OPERATION

ADAAG

> **4.11.3 Entrance. If platform lifts are used then they shall facilitate unassisted entry, operation, and exit from the lift in compliance with 4.11.2.**

ICC/ANSI A117.1-1998

The requirements are essentially similar plus the lift **shall not be attendant-operated.** The following additional requirements are also applicable.

> 408.2 Door and Gates. Lifts shall have low energy power-operated doors or gates complying with Section 404.3. Doors and gates shall remain open for 20 sec. minimum. End doors shall be 32 inches (815 mm) minimum clear width. Side doors shall be 42 inches (1065 mm) minimum clear width.
>
> > EXCEPTION: *Lifts having doors or gates on opposite sides shall be permitted to have manual doors or gates.*

CABO/ANSI A117.1-1992

> The requirements are essentially identical.

ANSI A117.1-1986

> There are no similar requirements.

UFAS

> The requirement is identical

OVERVIEW

> This requirement prohibits the use of attendant-operated wheelchair lifts. ASME A17.1 requires some wheelchair lift designs to be attendant-operated. Typically, inclined wheelchair lifts with folding platforms have platform guards complying with the requirements of ASME A17.1, Rule 2001.6c(2). Wheelchair lifts conforming to that requirement are required by ASME A17.1, Rule 2001.6c to be attendant-operated. These designs are not acceptable under ADAAG, CABO/ANSI A117.1-1992 or UFAS. The previous requirement in § 4.11.2 specified that wheelchair lifts must comply with ASME A17.1 Part XX.

> The prohibition of "attendant operation" should not be confused with "key operation" which is required by ASME A17.1 and all the referenced accessibility standards, which require wheelchair lifts to comply with ASME A17.1. The need for a key to operate the lift is no different than the need for a key to access a facility. Keyes are not prohibited by ADAAG.

4.3 ANALYSIS OF ESCALATOR REGULATIONS

The following provisions appear only in ADAAG for transportation facilities:

10.3.1 New Construction. New stations in rapid rail, light rail, commuter rail, intercity bus, intercity rail, high speed rail, and other fixed guideway systems (e.g. automated guideway transit, monorails, etc.) shall comply with the following provisions as applicable:

> **(16) Where provided in below grade stations, escalators shall have a minimum clear width of 32 inches. At the top and bottom of each escalator run, at least two contiguous treads shall be level beyond the comb plate before the riser begin to form. All**

> **escalators treads shall be marked by a strip of clearly contrasting color, 2 inches in width, placed parallel to and on the nose of each stop. The strip shall be of a material that is at least as slip resistant as the remainder of the tread. The edge of the tread shall be apparent from both ascending and descending directions.**

This provision in ADAAG is enforced by the DOT. The DOT regulations unlike the DOJ regulation have provisions to obtain approval of "equivalent facilitation." For the following reasons, DOT recognition of equivalent facilitation might be the prudent course to pursue. The ASME A17.1 definition of flat step is used to determine if two contiguous treads are level beyond the comb. The ASME A17.1 definition of flat step is as follows:

"The distance expressed in step lengths, the leading edge of the escalator step travels after emerging from the comb before moving vertically."

See Chapter 5 for the Access Board's technical assistance on § 10.3.1(16).

The 2 in. wide strip placed at the nose of each step could adversely affect the safety of the passengers, while attempting to make the step demarcation obvious to people with impaired vision. The same effect can be accomplished by placing the colored strip at the back, or heel, of the step without compromising overall safety.

The nose of the step is the part subject to the most severe wear. It is essential that it resists the effects of wear and maintains its dimensional stability to avoid objects from becoming caught between the steps or as they enter the combs.

Colored plastic strips can be filled or coated with a friction producing material, but it will not be comparable with the base step material in terms of wear resistance. A painted strip would, at best, be a temporary marking whose life could be measured in days at a public transportation facility. Accessibility would be compromised with repeated shutdowns for repainting.

The nose of an escalator step is a critical area from both a structural and dimensional perspective. Any modification to accommodate an insert at this area would require a complete redesign of the step. This is a lengthy process and one that would complicate the step design and the process to produce an accurate and reliable part. The differing physical properties of the base step material and the insert could lead to increased breakage and other damage at this critically exposed area.

The requirements also call for the colored strip to extend into the riser to be visible from both directions. The escalator step is not designed for walking up or down on. Its dimensions are to facilitate a safe transition from the stationary floor to the moving steps. These dimensions do not conform to the standards of stationary stairs. A colored strip at the heel of the step would provide an obvious contrast between adjacent steps at both upper and lower landings when boarding the escalator. The color on the nose would be hidden at these locations, and would tend to cause a distraction as the riser formed, drawing the eyes of boarding passengers away from the step they are attempting to board.

This series of colored lines moving up or down, away from the boarding passenger can produce disorientation and an uncomfortable feeling in certain individuals with normal vision. This may be exacerbated in individuals with visual impairments. This problem can be avoided when the strips are placed at the heel of the step. There, they are only visible during boarding and are removed from sight by step's geometry while the standing passenger is transported by the escalator along the incline.

4.4 DESTINATION ORIENTED ELEVATORS

In this type of system, the desired floor is selected prior to boarding the elevator. Several locations are usually provided in the lobby or approach to the lobby to enter the desired floor number. For each entry, there will be an indication of which elevator car to enter such as Car "A" or Car "B." The passenger then needs only to wait in front of the designated car and enter when it arrives. When the passenger boards, there will be car indicators, which provide information on which floors the car will stop. In many cases, where sufficient space is available, these indicators will be placed on the car jambs. When the car reaches the desired floor, the indicator will extinguish and a verbal announcement such as "Floor Fifteen" will be made so the passenger knows it is time to exit the car.

4.4.1 CALL BUTTONS

Call buttons in the lobby and halls approaching the lobby are generally provided in the form of a large keypad. Keypads are in standard telephone format and have black keys 19 mm minimum with 16 mm minimum visual white letters or symbols, and a 3 mm dia tactile dot on the number 5 key. The star appearing on the lower left key will be visually the same as the 5 pointed star in Table 407.2.11.2 and, when pushed, will enter a call to the "Main" floor designated for the elevator regardless of the floor number. The "Pound" key position in the lower right hand position is used to enter minus "-" floors such as usually provided for parking; however, instead of floors being "B1", "B2" or some other alpha numeric indication, they will be designated "-1", "-2" etc. The "Pound" key position and then a number in succession will enter the appropriate call. To enter floor call for higher floors such as floor fifteen, it is only necessary to push "1" and then "5" in succession.

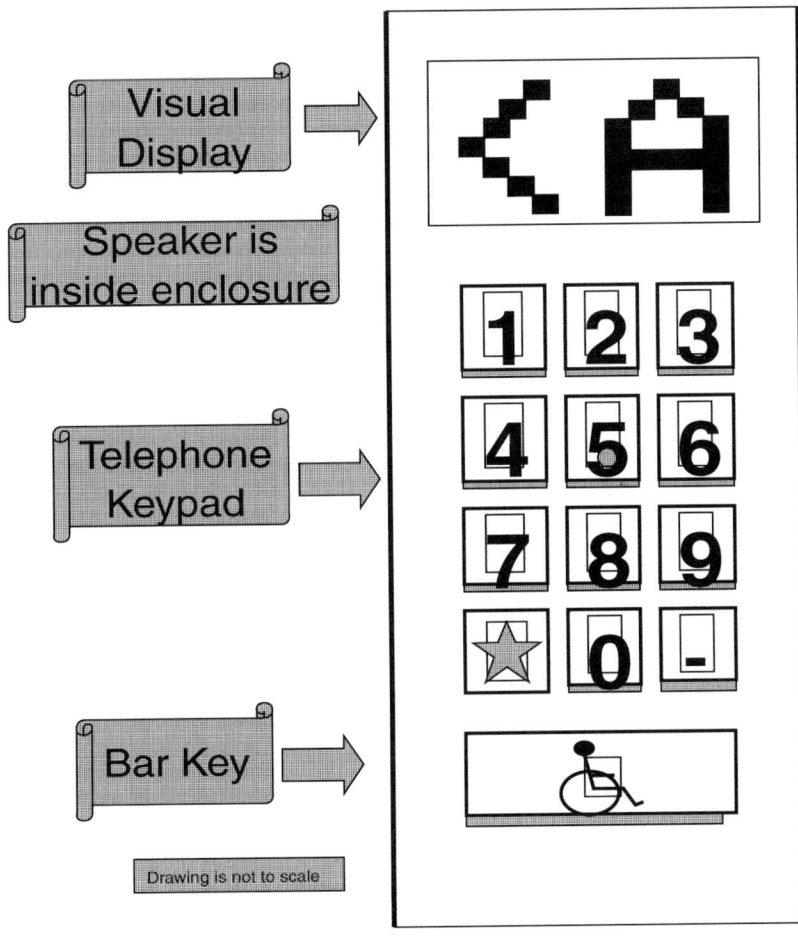

Courtesy George Kappenhagen, Schindler Elevator Corporation

4.4.2 HALL SIGNALS

Some hall signals are associated with the keypad itself and other hall signals are associated with the elevator position.

The keypad itself is provided with several items: a visual display located above the keypad, a bar key located below the keypad and a speaker located within the keypad enclosure. The visual display above the keypad provides the car designation "A", "B" etc. and several symbols to indicate the direction of car to enter: left, right, etc. The large bar key located below the keypad serves to turn on several accessibility features such as lobby audible signals, and more time to push successive keypad numbers. For those requiring such features, the bar key must be pushed before entering the floor number. The lobby audible signals cannot be left enabled all the time. Although individual signals are easy to follow, a very large number of signals initiated at multiple keypads in rapid succession are not individually distinguishable by those who need the signals. Following bay key activation and entry of a floor destination, a verbal announcement confirming the floor number and designation of which car to take will occur. The keypad speaker will then emit a unique audible tone, assigned to the specific car. A speaker at the car will emit the same tone and repeat it at intervals for the duration of the notification time. The duration will correspond to the distance from the particular keypad in question to a point directly in front of the specified car. Supplementary verbal announcements are also permitted, example: "there is no floor thirteen" but not required.

4.4.3 TACTILE CHARACTERS ON HOISTWAY ENTRANCES

As further correlation that the specified car has been reached, an elevator car identification is placed on both jambs immediately below the floor designation. As a further help, the floor designation is always numeric to match the keypad and the car designation is always alphabetic. The five pointed raised star to indicate the main entry is next to the main floor designation. In most cases, the "main" entry will be floor "1," however, some new building are adopting the convention used in most other countries and calling it floor "0." Then one floor up is "1" and one floor down is "-1." In some buildings with multiple banks of elevators exiting of different streets and having sloping terrain or other conditions, the "Main" floor may differ for each bank of elevators.

4.4.4 CAR CONTROLS

Except for the emergency controls and emergency communications, there are not any accessible control elements within the car since all floor destinations have already been entered before boarding the elevator.

4.4.5. CAR DISPLAYS

The in-car destination display will remain lit until the floor designated has been reached. The visible indication is extinguished after the call has been answered. Other enhancements such as supplementary verbal announcement and having the next designated stop on the destination display begin to flash are permitted but not required. The destination display should not be confused with the car position indicator, which is a different display.

4.5 LIMITED-USE/LIMITED-APPLICATION (LU/LA) ELEVATORS

LU/LA elevators are recognized in ICC/ANSI A117.1-1998. At the time this sec. edition of this Handbook is being written, they can be installed and considered accessible in jurisdictions enforcing the 1999 edition of the NBC and SBC and the facility does not have to comply with ADA. Examples include churches and private clubs. They cannot be installed in any facility that is required to comply with ADA. In new construction ADAAG §4.1.3(5), Exception (1)(b) states "if a building or facility is eligible for exemption (no elevator required) but a passenger elevator is nonetheless planned, the elevator shall meet the requirements of §4.10...." A LU/LA elevator cannot comply with ADAAG §4.10. In an existing building, an exemption is given to the above requirement for the inside car dimension. See ADAAG §4.1.6(3)(c)(ii) and (iii).

(ii) Where existing shaft configuration or technical infeasibility prohibits strict compliance with 4.10.9, the minimum car plan dimensions may be reduced by the minimum amount necessary, but in no case shall the inside car area be smaller than 48 in. by 48 in. (1220 mm by 1220 mm).

(iii) Equivalent facilitation may be provided with an elevator car of different dimensions when usability can be demonstrated and when all other elements required to be accessible comply with the applicable provisions of 4.10. For example, an elevator of 47 in. by 69 in. (1195 mm by 1755 mm) with a door opening on the narrow dimension could accommodate the standard wheelchair clearances shown in Fig. 4.

Conceivably, a LU/LA elevator could be provided in an existing facility. However, all of the requirements in ADAAG Section 4.10 (automatic operation, automatic doors, etc.) would apply except the requirement for car inside dimensions.

Safety requirements for LU/LA elevators were developed by the ASME A17 Main Committee in response to the demand for a low-rise, low cost alternative to a "full passenger elevator." In buildings and facilities where the latter is not required, it was felt that the LU/LA would provide a practical alternative for accessibility where a building owner may otherwise opt for no vertical access.

Private residence elevators and fully enclosed wheelchair lifts have been used to provide access in small buildings, though such applications violate ADAAG, ASME and the building code.

Many of the accessibility requirements for LU/LA elevators are the same as required for a new passenger elevator. However, the LU/LA elevator requirement recognize that it may not be economically practical to comply with all the requirements and achieve the objective of obtaining access where it may otherwise not be provided.

The rational for some of those requirements, which differ from that required for new passenger elevator are as follows.

Horizontal sliding doors must conform to all new elevator requirements, including the protective and reopening device requirements. Swing doors are permitted, but must be power operated and must remain open a full 20 sec. to compensate for the lack of protective/reopening equipment (that allows doors to begin closing after 3 sec.). Swinging doors must also be low energy power-operated.

Door and signal timing for hall calls (§407.2.6) and door delay for car calls (§407.2.7) are not necessary for LU/LA as persons will be immediately in front of the single car entrance when the call is initiated and the car responds. These sections assume a large elevator lobby in which someone must move to an answering car some distance from the call buttons, which he/she pushed.

LU/LA elevators will not provide space to maneuver within the car; the reference to 407.2.8 is derived from the space requirements in 408 Platform (wheelchair) lifts. Car doors must be positioned at the narrow ends of the clear floor space to ensure an accessible route into the elevator. Where a single car door requires backing into or out of the car, the 36 in. (915 mm) width provides the space required for maneuvering into or out of an alcove.

Car controls are required to be located on a sidewall to insure they are within reach.

Car position indicators (§407.2.12) are not necessary because the car control button will extinguish when the desired floor is reached. Audible indicators are not required as LU/LA elevators provide limited travel to only one or two floors where doorjamb markings are provided.

4.6 PRIVATE RESIDENCE

Requirements for accessible private residence elevators (ASME A17.1-1996, Part V) and private residence wheelchair lifts (ASME A17.1-1996, Part XXI) can be found in ICC/ANSI A117.1-1998, Chapter 10. Chapter 10 Section 1003 of this standard provides technical criteria for Type B dwelling units. These criteria are intended to be consistent with the intent of only the technical requirements of the Federal Fair Housing Amendments Act Accessibility Guidelines. These Type B dwelling units are intended to supplement, not replace, accessible Type A dwelling units as specified in Section 1002. The type dwelling unit will be specified in the building code. The building transportation industry should not concern itself with the type of dwelling unit. The requirements for private residence elevators and private residence wheelchair lifts are the same for Type A and Type B dwelling units.

4.7 MAINTENANCE

Suppose a facility is required by ADA to provide an elevator, and does provide it, but the elevator is frequently out of operation. What remedies does a disabled person have for the failure to keep the elevator operable?

At the limit, failure to keep an elevator operable is a failure to provide an elevator. If the elevator is never capable of providing service, it is not functionally an elevator. The DOJ regulations accept this reasoning. 28 C.F.R. § 36.211 requires that public accommodations maintain in operable working condition features of facilities and equipment that must be readily accessible and usable by the disabled, excepting only isolated or temporary interruptions in service due to maintenance or repairs.

If the elevator is operable some of the time, when does the level of unavailability amount to a breach of the duty to provide elevator service? Second, what is the appropriate remedy for such a breach? The answer to these questions may depend to some extent on whether the duty to provide the elevator arises under Title II or Title III.

Title III's remedies are primarily injunctive, with civil penalties and monetary damages in limited circumstances. Injunctive relief could include fairly detailed supervision of the allocation of resources to maintain the elevator, possibly under the auspices of a court-appointed receiver. A receiver would not be appointed, nor would a court supervise allocation of resources, however, unless the responsible entity demonstrated an inability or unwillingness to comply with a simple order to keep the elevator in operable condition.

If the Justice Department sues, ADA § 308(b)(2), 42 U.S.C. § 12188(b)(2), authorizes a court also to award monetary damages (excluding punitive damages) as persons aggrieved, but only when requested by the Justice Department. The same section authorizes civil penalties of US$50,000 for first violations and US$100,000 for each subsequent violation, "to vindicate the public interest." In addition, § 204(a) of the Civil Rights Act of 1964, 42 U.S.C. § 2000a-3(a), prohibiting discrimination in public accommodations, authorizes injunctions and other preventive relief and attorneys' fees.

ADAAG QUESTIONS & ANSWERS

5.

5. ADAAG QUESTIONS AND ANSWERS

The following questions and answers are arranged by ADAAG Section number. On June 8, 1992, representatives of NEII met with the ATBCB staff to obtain answers to numerous questions raised by the industry. The questions answered at that meeting are identified as **[ACCESS BOARD TECHNICAL ASSISTANCE]**.

Access Board Technical Assistance is intended solely as informal guidance and is not a determination of ones' legal rights or responsibilities under ADA, nor is it binding on the ATBCB or DOJ. However, it should be acceptable evidence of a good faith effort by the elevator industry to clarify the requirement in ADAAG.

§ 4.10 ELEVATORS
§ 4.10.1 GENERAL

Question 1: Do elevators undergoing alterations have to comply with all sections of the ASME A17.1-1990 Safety Code?
Answer 1: No, only the applicable Rules in Part X and Part XII of ASME A17.1-1990. Part X applies to inspections and tests. Part XII applies to alterations, repairs, replacement of parts and maintenance.

Question 2: Should an issue be made out of the gap between the floor sill and the hall sill? On several occasions, we received measurements, as much as 1¾ in. and according to ADAAG, it is not supposed to exceed 1¼ in.
Answer 2: Clearances between the car-platform sill and the hoistway edge of any landing sill have to be in compliance with ASME A17.1-1990, Section 108 for new installations. Alterations and modernization have to comply with ASME A17.1, Part XII including ASME A17.3.

Question 3: Will there be such a thing as a variance? If so, what is the procedure?
Answer 3: There are no procedures to obtain a variance.

Question 4: ADAAG references the ASME A17.1-1990 Code. Does this now mean all elevators must comply with Firefighters' Service Phase I and II per the 1990 Code, no matter what version the state or local government has adopted?
Answer 4: Unless the Justice Department rules that local codes meet or exceed the requirements of Title III of the ADA, the requirements of ASME A17.1-1990 apply.

On modernizations', ASME A17.1-1990, Part XII applies. The extent of modernization will determine if the A17.1-1990 version of Firefighters' Service applies.

ASME A17.1, Rule 1200.1 requires elevators that are altered to comply with ASME A17.3 as a minimum.
Note: Also see Question and Answer 7.

Question 5: We find no reference at all to handrails in the elevator cab. Has this been missed or is it not an issue? Are handrails required? If required, at what height are they set? Do they have to be on all sides or just one side?
Answer 5: Handrails are not required.

Question 6: Section 4.10.1 requires elevators to comply with ASME A17.1-1990. New installations have no problems meeting this requirement. Existing installations typically comply with the applicable code at the time of installation. When altered, ASME A17.1 Part XII specifies what requirements must be adhered to. Generally only that portion of the elevator that is altered must comply with ASME A17.1-1990. Does an existing elevator that complies with all the requirements in Section 4.10, with the exception of ASME A17.1-1990, comply with ADAAG?
Answer 6: Yes, if the elevator in question complies with ASME A17.1, Part XII.
[ACCESS BOARD TECHNICAL ASSISTANCE]

Question 7: A new addendum to ASME A17.1-1990 has recently been published. New addendum or editions are published yearly. Does an elevator that complies with an edition of ASME A17.1 later than 1990 and the requirements in ADAAG comply?
Answer 7: Compliance with ASME A17.1-1990 is mandatory, but more stringent requirements in later supplements or editions of A17.1 are acceptable. **[ACCESS BOARD TECHNICAL ASSISTANCE]**

Question 8: Section 4.10.1 requires mandatory compliance with the ASME A17.1-1990 Safety Code for Elevators and Escalators. Do states, such as California and Pennsylvania, who have different elevator and escalator safety codes now have to comply with ASME A17.1-1990?
Answer 8: Compliance with ASME A17.1-1990 is mandatory, but more stringent requirements are acceptable. Both local and state jurisdictions have the option of applying for certification of their codes to the DOJ. If it is determined by DOJ that the local codes meet or exceed ADAAG requirements, then compliance with local codes is acceptable.
[ACCESS BOARD TECHNICAL ASSISTANCE]

§ 4.10.2 AUTOMATIC OPERATION

Question 9: Is an elevator considered "automatic" if an attendant is in the elevator operating a car switch?
Answer 9: No. See definition for "operation, automatic" and "operation, car-switch" in ASME A17.1-1990.

Question 10: Is an elevator considered to be "self-leveling" if an attendant is in the elevator operating a car switch and stopping the elevator?
Answer 10: No.

Question 11: Does a "single automatic push button elevator" comply with 4.10.2?
Answer 11: Yes. "Single Automatic Operation" as defined in ASME A17.1 meets the ADAAG requirements. However, it is not clear how "single automatic operation" could be made to comply with all the requirements in ADAAG Section 4.10.

Question 12: Some elevator designs are not arranged to level or relevel with the door(s) opened. They are arranged such that the door(s) will not open until the car is level with the floor landing as specified in this requirement. Does this comply with Section 4.10.2?
Answer 12: Yes. The ADAAG requirements only address leveling. Releveling is not a requirement. **[ACCESS BOARD TECHNICAL ASSISTANCE]** Note: Also see Chapter 4, Section 4.1.2.

§ 4.10.3 HALL CALL BUTTONS

Question 13: If an entire hall button fixture (button and faceplate) is recessed 1 in. into the wall but the button is raised or flush to the faceplate does it comply?
Answer 13: No, it does not.

Question 14: Sections 4.10.3 and 4.10.12(1) requires the use of raised or flush buttons. Buttons may be provided with a ferrule (trim or luminous ring) that could be flush or raised from the surface of the cover plate. Attachment #1 illustrates a number of raised or flush designs. Please indicate which illustration(s) complies with ADAAG?
Answer 14: Only arrangements A and D comply. Arrangements B, C and E do not comply. **[ACCESS BOARD TECHNICAL ASSISTANCE]**

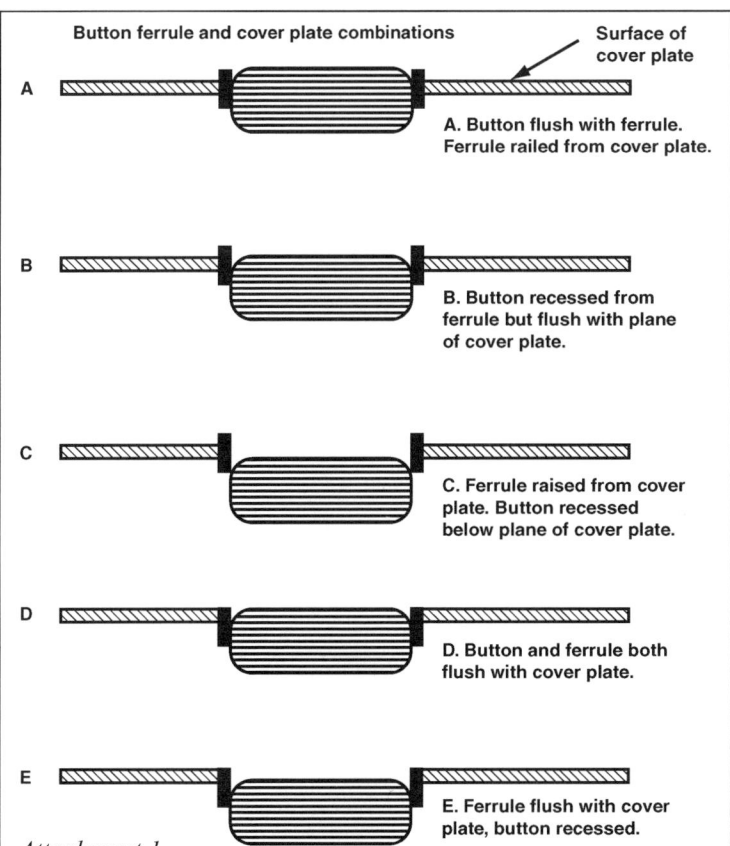

Attachment 1

Question 15: In a five car group of elevators, one of the cars can be removed from the group and operates off a separate riser of buttons. Does the separate button riser have to meet the heights requirements if the group riser and the car controls meet the requirement of 4.10.3 and 4.10.12(3)?
Answer 15: Yes.

Question 16: If there are two (2) risers of hall buttons in the same elevator lobby, are both required to meet the 42 in. requirement specified in 4.10.3?
Answer 16: Yes.

Question 17: ADAAG Fig. 20 shows that the centerline of an up/down hall button fixture is to be 42 in. The local code requires that the height from the floor to the centerline of the uppermost button shall be 42 in. maximum. Which regulations are applicable?
Answer 17: Compliance with 4.10.3 is mandatory, but more stringent requirements are acceptable. Both local and state jurisdictions have the option of applying for certification of their codes to the DOJ. If it is determined by DOJ that the local codes meet or exceed ADAAG requirements, then compliance with local codes is acceptable.

Question 18: How many in-car lanterns are required per car entrance? In some areas it has been sufficient to have one in each jamb for center-opening doors and one on the strike jamb side for 1- or 2-speed side slide doors.
Answer 18: A single in-car lantern will suffice as long as it is visible from the vicinity of the hall button(s).

Question 19: 4.10.4(2) states the visual elements of hall lanterns shall be at least 2½ in. in the smallest dimension. What does smallest dimension mean? Take the example of a chevron-type up arrow on a hall lantern. Which dimension is considered the smallest dimension?
Answer 19: Both the horizontal and vertical dimensions have to be at least 2½ in.

Question 20: Can we use horizontal mounted hall lanterns if the 72 in. requirement is met?
Answer 20: Yes.

§ 4.10.6 DOOR PROTECTION AND REOPENING DEVICE

Question 21: (1) Does this requirement establish a 20 second dwell time for all door operations?
(2) Can door time be cut out after initiation of a car call? (Ref. 4.10.8)
Answer 21: (1) No. The 20 second applies to the time the door reopening device must remain operational before nudging can be activated.
(2) No. Door open times have to comply with Sections 4.10.7 and 4.10.8.

Question 22: How many beams does a "light ray" device require?
Answer 22: Light rays require at least two beams, 5 in. and 29 in. above the sill.

Question 23: Section 4.10.6 requires a door protective device that will automatically stop and reopen the door without requiring contact with an obstruction. On an existing automatic elevator, Section 4.1.6(3)(c)(i) states, "if safety door edges are provided, the automatic door reopening device may be omitted (see 4.10.6)." A safety door edge typically stops and

reverses the door after contact with an obstruction. Is it correct that contact is thus permitted by the requirements in Section 4.1.6(3)(c)(i)?
Answer 23: Yes. [ACCESS BOARD TECHNICAL ASSISTANCE]

Question 24: A customer has requested we remove the limited door reversal feature from their elevators because it does not comply with Section 4.10.6. Is this correct?
Answer 24: Section 4.10.6 states that the reopening device "will stop and reopen a car door and hoistway door...., if the door becomes obstructed by an object or person." Section 4.10.6 does not say that the door should reopen fully.

Question 25: Must doors remain open for 20 sec. after reopening?
Answer 25: No. The requirement is that the "door reopening device ... remain effective for at least 20 sec." It does not say that the doors have to "remain open for 20 sec after reopening."

§ 4.10.7 DOOR AND SIGNAL TIMING FOR HALL CALLS

Question 26: Does the minimum acceptable notification time of 5 sec apply to both hall lanterns and in-car lanterns?
Answer 26: Yes.

Question 27: (1) Does "timing" start when the doors are fully opened or when the hall lantern gives a visible signal that the elevator is responding to the call?
(2) Can the door close button shorten door open time?
Answer 27: (1) Timing begins when the lantern is visible from the vicinity of the hall call buttons and audible signal is sounded.
(2) There appears to be no prohibition of direct passenger intervention, such as by the use of door close button to shorten the door open time.

Question 28: If the "signal time" is adequate without passenger transfer, is it adequate if the monitor (light ray) immediately stops dwell time when the beams are remade? This is also true for car call dwell time. Example: A passenger enters the car and the doors automatically start to close as soon as they are out of the door path thus reducing the door open time.
Answer 28: No. Indirect (e.g., automatic) passenger intervention, such as monitor operation, cannot shorten the door open time.

§ 4.10.8 DOOR DELAY FOR CAR CALLS

Question 29: If the "signal time" is adequate without passenger transfer, is it adequate if the monitor (light ray) immediately stops dwell time when the beams are remade? This is also true for car call dwell time. Example: Passengers enter the car and the doors automatically start to close as soon as they are out of the door path thus reducing the door open time.
Answer 29: No. Indirect (e.g., automatic) passenger intervention, such as monitor

operation, cannot shorten the door open time.

§ 4.10.9 FLOOR PLAN OF ELEVATOR CARS

Question 30: Some hospital size cars only have 64 in. wide dimension (4.10.9 requires 68 in. minimum width). If inside dimensions provide 60 in. wide circle for turning of a wheelchair, does it meet the intent of 4.10.9?
Answer 30: Although elevator cars, with a clear floor area in which a 60 in. diameter circle can be inscribed, would appear to meet the intent of the equivalent facilitation requirements of Section 2.2, the ATBCB has stated on the record that cab designs narrower then 68 in. are not acceptable except in transportation facilities covered by ADAAG Section 10.

§ 4.10.12 CAR CONTROLS

Question 31: (1) In the case of two operating panels in a car, do they both have to meet ADAAG requirements?
(2) In the case of a car with center-opening doors and one car operating panel, does it matter which return the car operating panel is in?
Answer 31: (1) Yes.
(2) No.

Question 32: Does a key operated button or keyless entry system comply with 4.10.12(1)?
Answer 32: Keys and car readers are permitted provided they meet the "operation" requirements of 4.27.4 and the "height" requirements of 4.10.12(3). **[ACCESS BOARD TECHNICAL ASSISTANCE]**

The key or card that is required to operate such a device in not regulated by ADAAG as it is considered a personal device. An individual who may need a specially modified key (i.e., large handle) or card could have that modification made or could request same as a "reasonable accommodation" under Title I of ADA.

Question 33: A4.10.12 is suggesting industry wide standardization of COPs. Any activity going on here?
Answer 33: Common issues are being addressed by NEII (e.g., COP location and height, Grade 2 Braille, etc.).

Question 34: What is Grade 2 Braille?
Answer 34: Grade 2 Braille consists of 250 contractions that are used in the printing of all books and magazines for the blind.

Question 35: What is the definition of "pictogram"?
Answer 35: There is no definition for "pictogram."

Question 36: Are the car button symbols pictograms?
Answer 36: No.

Question 37: Are mechanical buttons acceptable as long as they have a jewel light to the side of each button?
Answer 37: Yes, as long as the floor control buttons are raised or flush.

Question 38: Do button identification symbols (door open, emergency alarm, etc.) have to be identified by Braille as well as the equivalent verbal description? (Section 4.30.3)
Answer 38: The equivalent verbal description is not required. Braille is required.

Question 39: Are the door open and close buttons (front and rear) that are normally found in car operating panels considered "emergency controls"?
Answer 39: No, the door open and close buttons are not the emergency controls referenced in 4.10.12. However, the door open and close buttons should be grouped with the emergency control buttons as shown in Fig. 23(a).

Question 40: If an elevator has a floor plan in compliance with requirements in Section 4.10.9 and the car control panels are installed in compliance with the requirements of Section 4.10.12 as shown in Fig. 23, are we correct in assuming that the car control panels are acceptable for a side reach?
Answer 40: The dimensional requirements in Sections 4.2.5 and 4.2.6 determine if it's a front or side approach panel. Appendices A4.2.5 and A4.2.6 further clarify the "reach" ranges for persons seated in wheelchairs. **[ACCESS BOARD TECHNICAL ASSISTANCE]**

Question 41: How shall we define what constitutes a "side" or "front" approach?
Answer 41: Typically the car operating panel in the return panel shown in Fig. 23(d) and those shown in Fig. 23(c) are front approach panels (i.e., floor buttons shall be no higher than 48 in. above the finished floor.) Only the "alternate" car operating panel location in Fig. 23(d) is for side approach.
> **NOTE:** The dimensional requirements in 4.2.5 and 4.2.6 determine if it's a front or side approach, respectively. Also refer to A4.2.5 and A4.2.6 that further clarifies the "reach" ranges for persons seated in wheelchairs.

Question 42: Sections 4.10.3 and 4.10.12(1) requires the use of raised or flush buttons. Buttons may be provided with ferrule (trim or luminous ring) that could be flush or raised from the surface of the cover plate. The drawing shown in Question 14 illustrates a number of raised or flush designs. Please indicate which illustration(s) complies with ADAAG?
Answer 42: Only arrangements A and D comply. Arrangements B, C and E do not comply.

Question 43: Section 4.10.12(2) requires that all raised and Braille characters and symbols comply with 4.30.
(1) Fig. 23(a) specifies that the control button numeral height as 5/8 in. Are we correct that the character and number height must be as specified in Fig. 23(a) and not as specified in Section 4.30.3?
(2) Are we correct the symbols in Fig. 23(a) are not considered pictographs as specified in Section 4.30.4?
(3) Are the numerals (i.e., Arabic characters) and standard symbols in Fig. 23(a) required to be designated by Braille also? [e.g., the main entry floor is to be designated by a star as illustrated in Fig. 23(a)]
      (a) must the corresponding floor identification and Braille accompany the star; or
      (b) may a star be provided accompanied only with the Braille identification?
Answer 43: (1) Yes. The requirements of Section 4.30.3 only apply to signs over 80 in. from the floor.
(2) The pictograph requirements apply to building signage as specified in Section 4.1.3(16)(a).
(3) The standard numerals and symbols (e.g., door open button, door close button, alarm, emergency stop) are required to have Braille also. The in-car stop switch does not have to be identified with raised character or Braille. **[ACCESS BOARD TECHNICAL ASSISTANCE]**

Question 44: Is it necessary to have the Braille below the raised character on hoistway entrances or on car operating panels or can it be to one side (e.g., the left side)?
Answer 44: Sections 4.10.5 and 4.10.12 do not specify the location of the Braille.
    Note: ICC/ANSI A117.1-1998 Section 703.5.1 states that "Braille provided on elevator car controls shall be separated $^3/_{16}$ inch (4.8 mm) minimum either directly below or adjacent to the corresponding raised characters or symbols". See also Fig. B1 in this Handbook for required Braille description of car control symbols.

Question 45: To lower the top floor buttons to a maximum height of 48 in. and still maintain the emergency control buttons to a minimum height of 35 in., leaves only 13 in. to squeeze all the floor buttons within. This is not possible for elevators with more than 12-14 buttons. How are the requirements met, with the space limitation?
Answer 45: In order to meet the height limitations of 4.10.12(3), buttons will have to be rearranged (e.g., 3 across, etc.)

Question 46: Are mechanical buttons acceptable if there is a single up and down direction jewel that illuminates when a car button is registered?
Answer 46: No.

Question 47: Since the door open and door close buttons are not considered emergency controls, can they be located below the emergency stop (if required) and alarm button if these two are mounted on a 35 in. centerline from the finished floor?

Answer 47: No.

§ 4.10.13 CAR POSITION INDICATORS

Question 48: When an elevator is provided that only serves two stops;
(1) is a position indicator required, and
(2) is an audible floor passing/stopping signal required?
Answer 48: Yes to both parts of this question. **[ACCESS BOARD TECHNICAL ASSISTANCE]** Section 4.10.13 does not exempt two (2) stop elevators.

Question 49: Is 20 decibel (min) above ambient noise and at what distance is it measured?
Answer 49: Yes. The 20 decibel (min) is above ambient noise and it is measured inside the car enclosure. **[ACCESS BOARD TECHNICAL ASSISTANCE]**

Question 50: If an automatic verbal announcement of the floor number at which a car stops is provided, is an audible floor passing signal or verbal floor passing announcement also required?
Answer 50: No. **[ACCESS BOARD TECHNICAL ASSISTANCE]**

§ 4.10.14 EMERGENCY COMMUNICATIONS

Question 51: Does the installation of an emergency alarm button that is illuminated when the alarm is activated meet the intent of the last sentence in this Section?
Answer 51: Although an illuminated alarm button is desirable, it does not by itself meet the full intent of 4.10.14. In addition to the illuminated alarm button and the requirements of ASME A17.1-1990, Rule 211.1, an alarm acknowledge indicator (e.g., help is on the way) is also needed for persons with speech and/or hearing impairments. **[ACCESS BOARD TECHNICAL ASSISTANCE]**

Question 52: Is an emergency communication system required by 4.10.14 of the ADAAG?
Answer 52: Yes. ASME A17.1 Rule 211.1 requires emergency communication.

Question 53: The emergency communication system shall not require voice communication per 4.10.14. A voice only system is inaccessible to persons with speech or hearing impairments. It would appear this would preclude a telephone meeting this requirement. Any comments?
Answer 53: No. See response above.

Question 54: This Section says operable parts shall be identified by raised or recessed symbols or letters complying with 4.30. Do we need to add identification for the telephone cabinet and handset?
Answer 54: Yes.

Question 55: This Section states that the highest operable part of a two-way communication system shall be a maximum of 48 in. from the floor of the car. An intercom two-way speaker is normally higher than 48 in. Does this apply only to devices that require some physical operation?
Answer 55: Yes.

Question 56: Does a telephone require a raised symbol and Braille?
Answer 56: Yes.

Question 57: Does a hands-free phone in a telephone cabinet that (1) has a build in auto dialer (monitored 24 hours a day), (2) is capable of identifying the elevator and building location without voice communication and (3) has an indicator light that tells the hearing impaired that help is on the way, comply?
Answer 57: Yes, provided the phone is properly labeled with operating instruction.

§ 4.11 PLATFORM LIFTS (WHEELCHAIR LIFTS)
§ 4.11.2 OTHER REQUIREMENTS

Question 58: What is the minimum size platform that is required?
Answer 58: The minimum size platform is 30 in. by 48 in. as required by the cross reference to § 4.2.4.

§ 4.11.3 ENTRANCE

Question 59: Is key operation prohibited?
Answer 59: No. To the contrary, key operation is required. ADAAG § 4.11.2 requires wheelchair lifts to comply with ASME A17.1-1990. ASME A17.1-1990 requires that wheelchair lifts be key operated.

   Keys must meet the "operation" requirements of 4.27.4 and the "height" requirements of 4.2.5 and 4.2.6.

Question 60: Are "attendant-operated lifts" that meet the requirements of ASME A17.1-1990, Part XX permitted?
Answer 60: No. "Attendant-operated lifts" per ASME A17.1, Part XX do not permit "unassisted operation" as required by ADAAG § 4.11.3.

Question 61: Does an inclined wheelchair lift provided with 6 in. high sides in conformance with ASME A17.1-1990, Rule 2001.6c(2) comply with ADAAG?
Answer 61: No. ASME A17.1, Rule 2001.6c requires wheelchair lifts conforming to Rule 2001.6c to be attendant operated. Attendant operated wheelchair lifts do not permit "unassisted operation" as required by ADAAG § 4.11.3.

To comply with ADAAG a 42 in. high sides and a 42 in. high door is required on an inclined wheelchair lift. See ASME A17.1 Rule 2001.6c(1).

4.30 SIGNAGE

Question 62: What is the application of "A4.30 signage"?
Answer 62: The material in Appendix 4.30 is advisory only (e.g., the standard dimensions in A4.30.4 are not mandatory).

Question 63: What portion of 4.30.7 applies to elevators?
Answer 63: The requirements of Section 4.30.7 (Symbols of Accessibility) do not apply to the control button numerals or standard symbols as shown in Fig. 23(a) of Section 4.10.12(2). Accessibility symbols include, but are not limited to those in Fig. 43(a) and (b), bathroom signs, etc.

Question 64: (1) Is literary Braille (A4.30.4) the same as Grade 2 Braille?
(2) If the Response is "yes," do we have to comply with the standard dimensions for literary Braille?
Answer 64: (1) Yes.
(2) No. **[ACCESS BOARD TECHNICAL ASSISTANCE]**
Note: See Appendix B for ANSI A117.1-1992 signage and Braille requirements, that are more stringent than those in ADAAG.

§ 10.3.1(16) TRANSPORTATION (ESCALATORS)

Question 65: (1) Do the requirements of this Section only apply to new construction?
(2) Are there any ADAAG requirements for existing escalators in public or private office buildings, hotels or retail stores?
Answer 65: (1) Yes.
(2) No.

Question 66: This Section states "at least two contiguous treads shall be level beyond the combplate before the risers begin to form." ASME A17.1 defines a flat step as "the distance, expressed in step lengths, that the leading edge of the escalator step travels after emerging from the comb before moving vertically." The ASME A17.1 definition defines the escalator illustrated in Attachment #2, Fig. (a) as having two flat steps. Would the condition illustrated in Attachment #2, Fig. (a) comply with this requirement or must it be as illustrated in Attachment #2, Fig. (b)?

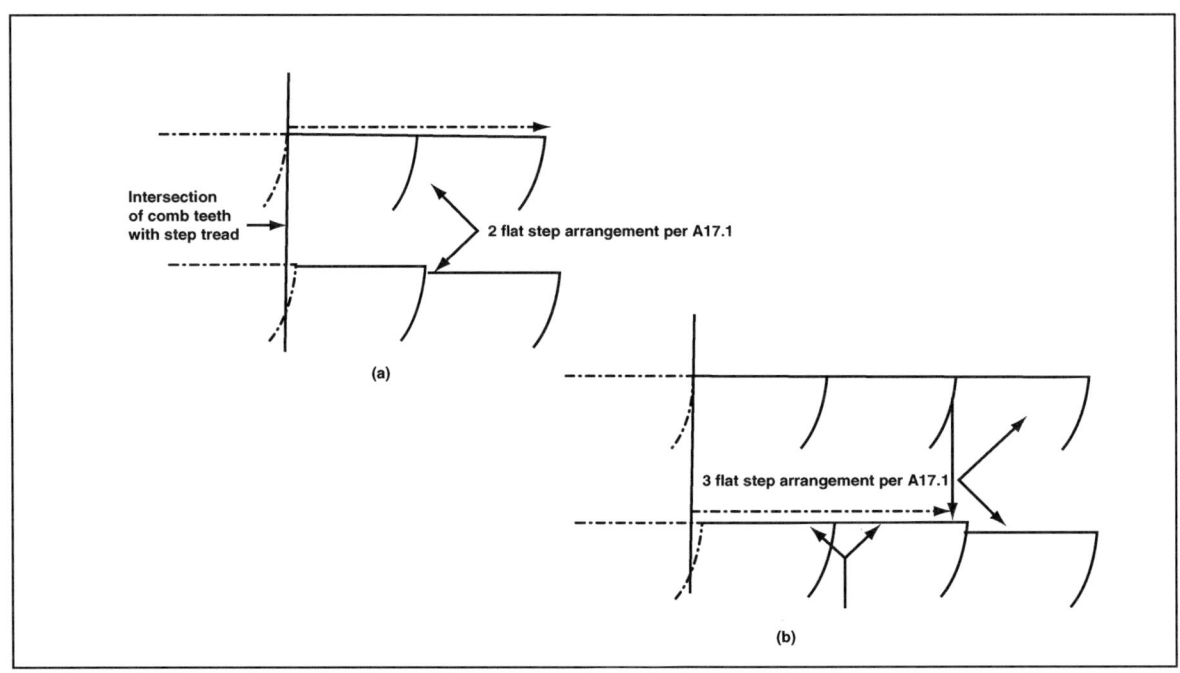

Attachment 2

Answer 66: The condition in Fig. (a) illustrates the requirements of this Section, but the condition in Fig. (b) is preferred. **[ACCESS BOARD TECHNICAL ASSISTANCE]**

10.3.1(17) TRANSPORTATION (ELEVATORS)

Question 67: Will vision panels, glass doors and/or glass enclosures meet the requirements of this section?

Answer 67: Yes, as long as they meet the requirements of the ASME A17.1-1990 Safety Code for Elevators and Escalators (e.g., Rules 110.7, 204.2e, 204.5i).

ACCESSIBILITY STANDARDS COMPARISON CHART

6.

VERTICAL TRANSPORTATION ACCESSIBILITY STANDARDS
Comparison Chart

14 December, 1999

ADAAG	ICC/ANSI A117.1 1998	CABO/ANSI A117.1-1992	ANSI A117.1-1986	UFAS
	105.2.3 Safety Code for Elevators and Escalators. ASME/ANSI A17.1-1996	**3.3 Referenced American National Standards.** The following American National Standards are referenced in this document. ASME/ANSI A17.1-1990, Safety Code for Elevators and Escalators (including Addenda ASME/ANSI A17.1a-1991)		
	106.5 Defined Terms **Destination oriented elevator system:** An elevator system that provides lobby controls to select destination floors, lobby indicators designating which elevator to board, and a car indicator designating the floors at which the car will stop.			
4.10 Elevators	**407 Elevators** **407.1 General.** Elevators required to be accessible shall comply with Section 407.2. Destination-oriented elevators required to be accessible shall comply with Section 407.3. Limited use/limited application elevators required to be accessible shall comply with Section 407.4. Altered elements of existing elevators shall comply with Section 407.5.	**4.10 Elevators**	**4.10 Elevators**	**4.10 Elevators**
		4.10.1 Elevators - New		
4.10.1 General. Accessible elevators shall be on an accessible route and shall comply with 4.10 and with the ASME A17.1-1990, Safety Code for Elevators and Escalators. Freight elevators shall not be considered as meeting the requirements of this section unless the only elevators provided are used as combination passenger and freight elevators for the public and employees.	**407.2 Elevators.** Elevators shall comply with Sections 407.2.1 through 407.2.13 and ASME/ANSI A17.1. They shall be passenger elevators.	**4.10.1.1 General.** Accessible passenger elevators shall comply with 4.10 and ASME/ANSI A17.1. Freight elevators shall not be considered as meeting the requirements of this section unless the only elevators provided are used as combination passenger and freight elevators.	**4.10.1 General.** Passenger elevators on accessible routes shall comply with ANSI/ASME A17.1-1984 and A17.1a-1985. This standard does not preclude the use of residential elevators or wheelchair lifts when appropriate and approved by administrative authorities. Freight elevators shall not be considered as meeting the requirements of this section unless the only elevators provided are used as combination passenger and freight elevators.	**4.10.1 General.** Accessible elevators shall be on an accessible route and shall comply with 4.10 and with the American National Standard Safety Code for Elevators, Dumbwaiters, Escalators, and Moving Walks ANSI A17.1-1978 and A17.1a-1979. This standard does not preclude the use of residential or fully enclosed wheelchair lifts when appropriate and approved by administrative authorities. Freight elevators shall not be considered as meeting the

VERTICAL TRANSPORTATION ACCESSIBILITY STANDARDS
Comparison Chart

ADAAG	ICC/ANSI A117.1 1998	CABO/ANSI A117.1-1992	ANSI A117.1-1986	UFAS
				requirements of this section, unless the only elevators provided are used as combination passenger and freight elevators for the public and employees.
4.10.2 Automatic Operation. Elevator operation shall be automatic. Each car shall be equipped with a self-leveling feature that will automatically bring the car to floor landings within a tolerance of 1/2 in (13 mm) under rated loading to zero loading conditions. This self-leveling feature shall be automatic and independent of the operating device and shall correct the overtravel or undertravel.	**407.2.1 Automatic Operation.** Elevator operation shall be automatic. Each car shall be equipped with a self-leveling feature that will automatically bring and maintain the car at floor landings within a tolerance of 1/2 inch (13 mm) under rated loading to zero loading conditions.	**4.10.1.2 Automatic Operations.** Elevator operation shall be automatic. Each car shall be equipped with a self-leveling feature that will automatically bring the car to floor landings within a tolerance of 1/2 in (13 mm) under rated loading to zero loading conditions. This self-leveling feature shall be automatic and independent of the operable part and shall correct for overtravel or undertravel.	**4.10.2 Automatic Operations.** Elevator operation shall be automatic. Each car shall be equipped with a self-leveling feature that will automatically bring the car to floor landings within a tolerance of 1/2 in (13 mm) under rated loading to zero loading conditions. This self-leveling feature shall be automatic and independent of the operating device and shall correct for overtravel or undertravel.	**4.10.2 Automatic Operations.** Elevator operation shall be automatic. Each car shall be equipped with a self-leveling feature that will automatically bring the car to floor landings within a tolerance of 1/2 in (13 mm) under rated loading to zero loading conditions. This self-leveling feature shall be automatic and independent of the operating device and shall correct the overtravel or undertravel.
4.10.3 Hall Call Buttons. Call buttons in elevator lobbies and halls shall be centered at 42 in (1065 mm) above the floor. Such call buttons shall have visual signals to indicate when each call is registered and when each call in answered. Call buttons shall be a minimum of 3/4 in (19 mm) in the smallest dimension. The button designating the up direction shall be on top. (see Fig. 20) Buttons shall be raised or flush. Objects mounted beneath hall call buttons shall not project into the elevator lobby more than 4 in (100 mm).	**407.2.2 Call Buttons.** Call buttons in elevator lobbies and halls shall be 35 inches (890 mm) minimum and 48 inches (1220 mm) maximum above the floor or ground, measured to the centerline of the buttons. A clear floor or ground space complying with Section 305 shall be provided. Such call buttons shall have visual signals to indicate when each call is registered and when each call is answered. Call buttons shall be 3/4 inch (19 mm) minimum in their smallest dimension. The button that designates the up direction shall be above the button that designates the down direction. Buttons shall be raised or flush. Objects beneath hall call buttons shall protrude 1 inch (25 mm) maximum.	**4.10.1.3 Call Buttons.** Call buttons in elevator lobbies and halls shall be centered at 42 in (1065 mm) above the floor. See Fig. B4.10.1. Such call buttons shall have visual signals to indicate when each call is registered and when each call is answered. Call buttons shall be 3/4 in (19 mm) minimum in the smallest dimension. The button that designates the up direction shall be located above the button that designates the down direction. Objects located beneath hall call buttons shall protrude into the elevator lobby 4 in (100 mm) maximum.	**4.10.3 Hall Call Buttons.** Call buttons in elevator lobbies and halls shall be centered at 42 in (1065 mm) above the floor. Such call buttons shall have visual signals to indicate when each call is registered and when each call is answered. Call buttons shall be a minimum of 3/4 in (19 mm) in the smallest dimension. The button designating the up direction shall be on top (see Fig. 20).	**4.10.3 Hall Call Buttons.** Call buttons in elevator lobbies and halls shall be centered at 42 in (1065 mm) above the floor. Such call buttons shall have visual signals to indicate when each call is registered and when each call is answered. Call buttons shall be a minimum of 3/4 in (19 mm) in the smallest dimension. The button designating the up direction shall be on top (see Fig. 20). Buttons shall be raised or flush. Objects mounted beneath hall call buttons shall not project into the elevator lobby more than 4 in (100 mm).
4.10.4 Hall Lanterns. A visible and audible signal shall be provided at each hoistway entrance to indicate which car is answering a call.	**407.2.3 Hall Signals.** A visible and audible signal shall be provided at each hoistway entrance to indicate which car is answering a call and the direction of travel, except that signals in cars, visible from the floor area adjacent to the hall call buttons, and complying with the requirements of this	**4.10.1.4 Hall Signals.** A visible and audible signal shall be provided at each hoistway entrance to indicate which car is answering a call and the direction of travel, except that in-car signals located in cars, visible from the floor area adjacent to the hall call buttons, and conforming to the	**4.10.4 Hall Lanterns.** A visible and audible signal shall be provided at each hoistway entrance to indicate which car is answering a call. Audible signals shall sound once for the up direction and twice for the down direction, or shall have verbal annunciators that say "up" or "down	**4.10.4 Hall Lanterns.** A visible and audible signal shall be provided at each hoistway entrance to indicate which car is answering a call. Audible signals shall sound once for the up direction and twice for the down direction, or shall have verbal annunciators that say "up" or "down."

VERTICAL TRANSPORTATION ACCESSIBILITY STANDARDS
Comparison Chart

ADAAG	ICC/ANSI A117.1 1998	CABO/ANSI A117.1-1992	ANSI A117.1-1986	UFAS
	subsection, shall be permitted.	requirements of this subsection, shall be acceptable.		
Audible signals shall sound once for the up direction and twice for the down direction or shall have verbal annunciators that say "up" or "down."	**407.2.3.1 Audible Signals.** Audible signals shall sound once for the up direction and twice for the down direction, or shall have verbal annunciators that state the word "up" or "down." Audible signals shall have a frequency of 1500 Hz maximum. The audible signal or verbal annunciator shall be 10 dBA minimum above ambient, but shall not exceed 80 dBA maximum, measured at the hall call button.	Audible signals shall sound once for the up direction and twice for the down direction, or shall have verbal annunciators that state the word "up" or "down."		
Visible signals shall have the following features:	**407.2.3.2 Visible Signals.** Visible signals shall comply with Sections 407.2.3.2.1 through 407.2.3.2.3.	Visible signals shall have the following features	" Visible signals shall have the following features:	Visible signals shall have the following features:
(1) Hall lantern fixtures shall be mounted so that their centerline is at least 72 in (1830 mm) above the lobby floor. (see Fig. 20.)	**407.2.3.2.1 Height.** Hall signal fixtures shall be 72 inches (1830 mm) minimum above the floor or ground, measured to the centerline of the fixture.	Hall signal fixtures shall be centered at 72 in (1830 mm) minimum above the lobby floor. See Fig. B4.10.1.	(1) Hall lantern fixtures shall be mounted so that their centerline is at least 72 in (1830 mm) above the lobby floor.	(1) Hall lantern fixtures shall be mounted so that their centerline is at least 72 in (1830 mm) above the lobby floor.
(2) Visual elements shall be at least 2-1/2 in (64 mm) in the smallest dimension.	**407.2.3.2.2 Size.** The visible signal elements shall be 2-1/2 inches (63 mm) minimum in their smallest dimension.	The visible signal elements shall be 2-1/2 in (63 mm) minimum in the smallest dimension.	(2) Visual elements shall be at least 2-1/2 in (63 mm) in the smallest dimension.	(2) Visual elements shall be at least 2-1/2 in (64 mm) in the smallest dimension.
(3) Signals shall be visible from the vicinity of the hall call button (see Fig. 20). In-car lanterns located in cars, visible from the vicinity of hall call buttons, and conforming to the above requirements, shall be acceptable.	**407.2.3.2.3 Visibility.** Signals shall be visible from the floor area adjacent to the hall call button.	Signals shall be visible from the floor area adjacent to the hall call button	(3) Signals shall be visible from the vicinity of the hall call button. In-car lanterns located in cars, visible from the vicinity of hall call buttons, and conforming to the above requirements, shall be acceptable (see Fig. 20).	(3) Signals shall be visible from the vicinity of the hall call button. In-car lanterns located in cars, visible from the vicinity of hall call buttons, and conforming to the above requirements, shall be acceptable (see Fig. 20).
4.10.5 Raised and Braille Characters on Hoistway Entrances. All elevator hoistway entrances shall have raised and Braille floor designations provided on both jambs. The centerline of the character shall be 60 in (1525 mm) above finish floor. Such characters shall be 2 in (50 mm) high and shall comply with 4.30.4. Permanently applied plates are acceptable if they are	**407.2.4 Tactile Characters on Hoistway Entrances.** Tactile character and Braille floor designations shall be provided on both jambs of elevator hoistway entrances and shall be 60 inches (1525 mm) above the floor or ground, measured from the baseline of the characters. A tactile star shall also be provided on both jambs at the main entry level. Such characters shall be 2 inches	**4.10.1.5* Tactile Signage on Hoistway Entrances.** Raised character and Braille floor designations shall be provided on both jambs of elevator hoistway entrances and shall be centered at 60 in (1525 mm) above the floor. See Fig. B4.10.1. Such characters shall be 2 in (51 mm) high nominal and shall comply with 4.28.6.	**4.10.5 Raised Characters on Hoistway Entrances.** All elevator hoistway entrances shall have raised floor designations provided on both jambs. The centerline of the characters shall be 60 in (1525 mm) from the floor. Such characters shall be a nominal 2 in (51 mm) in height (see 3.2) and shall comply with 4.28. Permanently applied plates are acceptable if they are	**4.10.5 Raised Character on Hoistway Entrances.** All elevator hoistway entrances shall have raised floor designations provided on both jambs. The centerline of the character shall be 60 in (1525 mm) from the floor. Such characters shall be 2 in (50 mm) high and shall comply with 4.30. Permanently applied plates are acceptable if they are permanently fixed to the

VERTICAL TRANSPORTATION ACCESSIBILITY STANDARDS
Comparison Chart

ADAAG	ICC/ANSI A117.1 1998	CABO/ANSI A117.1-1992	ANSI A117.1-1986	UFAS
permanently fixed to the jambs. (see Fig. 20.)	(51 mm) high and shall comply with Section 703.2.		permanently fixed to the jambs (see Fig. 20).	jambs. (See Fig. 20).
4.10.6* Door Protective and Reopening Device. Elevator doors shall open and close automatically. They shall be provided with a reopening device that will stop and reopen a car door and hoistway door automatically if the door becomes obstructed by an object or person. The device shall be capable of completing these operations without requiring contact for an obstruction passing through the opening at heights of 5 in and 29 in (125 mm and 735 mm) above finish floor (see Fig. 20). Door reopening devices shall remain effective for at least 20 seconds. After such interval, doors may close in accordance with the requirements of ASME A17.1-1990.	**407.2.5 Doors.** Elevator doors shall be the horizontal type. Elevator hoistway and car doors shall open and close automatically. Elevator doors shall be provided with a reopening device that shall stop and reopen a car door and hoistway door automatically if the door becomes obstructed by an object or person. The device shall be activated by sensing an obstruction passing through the door opening at 5 inches (125 mm) and at 29 inches (735 mm) above the floor or ground. The device shall not require physical contact to be activated, although contact may occur before the door reverses. Door reopening devices shall remain effective for 20 seconds minimum.	**4.10.1.6* Door Protective and Reopening Device.** Elevator doors shall open and close automatically. Elevator doors shall be provided with a reopening device that shall stop and reopen a car door and hoistway door automatically if the door becomes obstructed by an object or person. The device shall be activated by sensing an obstruction passing through the door opening at 5 in and at 29 in (125 mm and 735 mm) above the floor. The device shall not require physical contact to be activated, although contact may occur before the door reverses. Door reopening devices shall remain effective for 20 seconds minimum.	**4.10.6* Door Protective and Reopening Device.** Elevator doors shall open and close automatically. They shall be provided with a reopening device that will stop and reopen a car door and hoistway door automatically if the door becomes obstructed by an object or person. The device shall be activated by sensing an obstruction passing through the door between 5 in and 29 in (125 mm and 735 mm) above the floor. It shall not require physical contact to be activated, although contact may occur before the reverses (see Fig. 20). Door reopening devices shall remain effective for at least 20 seconds. After such interval, doors may close in accordance with the requirements of ANSI/ASME A17.1-1984 and A17.1a-1985.	**4.10.6* Door Protective and Reopening Device.** Elevator doors shall open and close automatically. They shall be provided with a reopening device that will stop and reopen a car door and hoistway door automatically if the door becomes obstructed by an object or person. The device shall be capable of completing these operations without requiring contact for an obstruction passing through the openings at heights of 5 in and 29 in (125 mm and 735 mm) from the floor. (see Fig. 20). Door reopening devices shall remain effective for at least 20 seconds. After such interval, doors may close in accordance with the requirements of ANSI A17.1-1978 and A17.1a-1979.
4.10.7* Door and Signal Timing for Hall Calls. The minimum acceptable time from notification that a car is answering a call until the doors of that car start to close shall be calculated from the following equation: $T = D/(1.5 \text{ ft/s})$ or $T = D/(445 \text{ mm/s})$ where T total time in seconds and D distance (in feet or millimeters) from a point in the lobby or corridor 60 in. (1525 mm) directly in front of the farthest call button controlling that car to the centerline of its hoistway door (see Fig. 21). For cars with in-car lanterns, T begins when the lantern is visible from the vicinity of the hall call buttons and an audible signal is sounded. The minimum acceptable notification time shall be 5 seconds.	**407.2.6 Door and Signal Timing for Hall Calls.** The minimum acceptable time from notification that a car is answering a call until the doors starts to close shall be calculated by the following equation, but shall not be less than 5 seconds: $T = D/1.5 \text{ ft/s} (D/445 \text{ mm/s})$ where T = total time in seconds and D = distance in feet (millimeters) from the point in the lobby or corridor 60 inches (1525 mm) directly in front of the farthest call button controlling that car to the centerline of its hoistway door. For cars with in-car signals, T begins when the signal is visible from the point 60 inches (1525 mm) directly in front of the farthest hall call button and the audible signal is sounded.	**4.10.1.7* Door and Signal Timing for Hall Calls.** The minimum acceptable time from notification that a car is answering a call until the doors of that car start to close shall be calculated from one of the following equations: $T = D/(1.5 \text{ ft/s})$ or $T = D/(445 \text{ mm/s})$ = 5 seconds minimum where T = total time in seconds and D = distance (in feet or millimeters) from the point in the lobby or corridor 60 in (1525 mm) directly in front of the farthest call button controlling that car to the centerline of its hoistway door. For cars with in-car signals, T begins when the signal is visible from the point 60 in (1525 mm) directly in front of the farthest	**4.10.7* Door and Signal Timing for Hall Calls.** The minimum acceptable time from notification that a car is answering a call until the doors of that car start to close shall be calculated from one of the following equations: $T = D/(1.5 \text{ ft/s})$ or $T = D/(445 \text{ mm/s})$ where T= total time in seconds and D= distance (in feet or millimeters) from a point in the lobby or corridor 60 in (1525 mm) directly in front of the farthest call button controlling that car to the centerline of its hoistway door (see Fig. 21). The minimum acceptable notification time shall be 5 seconds. For cars with in-car lanterns, T begins when the lantern is visible from the vicinity of the hall call buttons and an audible signal is	**4.10.7* Door and Signal Timing for Hall Calls.** The minimum acceptable time from notification that a car is answering a call until the doors of that car start to close shall be calculated from one of the following equations: $T = D/(1.5 \text{ ft/s})$ or $T = D/(445 \text{ mm/s})$ where T = total time in seconds and D = distance (in feet or millimeters) from a point in the lobby or corridor 60 in (1525 mm) directly in front of the farthest call button controlling that car to the centerline of its hoistway door (see Fig. 21). For cars with in-car lanterns, T begins when the lantern is visible from the vicinity of the hall call buttons and an audible signal is sounded. The minimum acceptable notification time shall be 5 seconds.

VERTICAL TRANSPORTATION ACCESSIBILITY STANDARDS
Comparison Chart

ADAAG	ICC/ANSI A117.1 1998	CABO/ANSI A117.1-1992	ANSI A117.1-1986	UFAS
		hall call button and the audible signal is sounded.	sounded.	
4.10.8 Door Delay for Car Calls. The minimum time for elevator doors to remain fully open in response to a car call shall be 3 seconds.	**407.2.7 Door Delay for Car Calls.** Elevator doors shall remain fully open in response to a car call for 3 seconds minimum.	**4.10.1.8 Door Delay for Car Calls.** Elevator doors shall remain fully open in response to a car call for 3 seconds minimum.	**4.10.8 Door Delay for Car Calls.** The minimum time for elevator doors to remain fully open in response to a car call shall be 3 seconds.	**4.10.8 Door Delay for Car Calls.** The minimum time for elevator doors to remain fully open in response to a car call shall be 3 seconds.
4.10.9 Floor Plan of Elevator Cars. The floor area of elevator cars shall provide space for wheelchair users to enter the car, maneuver within reach of controls, and exit from the car. Acceptable door opening and inside dimensions shall be as shown in Fig. 22. The clearance between the car platform sill and the edge of any hoistway landing shall be no greater than 1-1/4 in (32 mm).	**407.2.8 Inside Dimensions of Elevator Cars.** The clear width of elevator doors and the inside dimensions of elevator cars shall comply with Table 407.2.8.	**4.10.1.9* Inside Dimensions of Elevator Cars.** The inside dimensions of elevator cars shall provide space for wheelchair users to enter the car, maneuver within reach of controls, and exit from the car. The clearance between the car platform sill and the edge of any hoistway landing shall be 1 1/4 in (32 mm) maximum.	**4.10.9 Floor Plan of Elevator Cars.** The floor area of elevator cars shall provide space for wheelchair users to enter the car, maneuver within reach of controls, and exit from the car. Acceptable door opening and inside dimensions shall be as shown in Fig. 22. The clearance between the car platform sill and the edge of any hoistway landing shall be no greater than 1-1/4 in (32 mm).	**4.10.9 Floor Plan of Elevator Cars.** The floor area of elevator cars shall provide space for wheelchair users to enter the car, maneuver within reach of controls, and exit from the car. Acceptable door opening and inside dimensions shall be as shown in Fig. 22. The clearance between the car platform sill and the edge of any hoistway landing shall be no greater than 1-1/4 in (32 mm).
4.10.10 Floor Surfaces. Floor surfaces in elevator cars shall comply with 4.5.	**407.2.9 Floor Surfaces.** Floor surfaces in elevator cars shall comply with Section 302. The horizontal clearance between the edge of the car platform sill and the edge of the landing sill shall be 1-1/4 inches (32 mm) maximum.	**4.10.1.10 Floor Surfaces.** Floor surfaces in elevator cars shall comply with 4.5.	**4.10.10 Floor Surfaces.** Floor coverings shall comply with 4.5.	**4.10.10 Floor Surfaces.** Floor surfaces shall comply with 4.5.
4.10.11 Illumination Levels. The level of illumination at the car controls, platform, and car threshold and landing sill shall be at least 5 footcandles (53.8 lux).	**407.2.10 Illumination Levels.** The level of illumination at the car controls, platform, and car threshold and landing sill shall be 5 footcandles (54 lux) minimum.	**4.10.1.11 Illumination Levels.** The level of illumination at the car controls, platform, and car threshold and landing sill shall be 5 footcandles (53.8 lux) minimum.	**4.10.11 Illumination Levels.** The level of illumination at the car controls, platform, and car threshold and landing sill shall be at least 5 footcandles (53.8 lux).	**4.10.11 Illumination Levels.** The level of illumination at the car controls, platform, and car threshold and landing sill shall be at least 5 footcandles (53.8 lux).
4.10.12* Car Controls. Elevator control panels shall have the following features:	**407.2.11 Car Controls.** Elevator controls shall comply with Sections 407.2.11.1 through 407.2.11.4.	**4.10.1.12* Car Controls.** Elevator control panels shall have the following features:	**4.10.12* Car Controls.** Elevator control panels shall have the following features:	**4.10.12* Car Controls.** Elevator control panels shall have the following features:
(1) Buttons. All control buttons shall be at least 3/4 in (19 mm) in their smallest dimension. They shall be raised or flush.	**407.2.11.1 Buttons.** Buttons shall be 3/4 inch (19 mm) minimum in their smallest dimension. Buttons shall be raised or flush. Except where provided in a standard telephone keypad arrangement, buttons shall be arranged with numbers in ascending order. Where two or more columns of buttons are provided they shall read from left to right.	**4.10.1.12.1** Control buttons shall be 3/4 in (19 mm) minimum in their smallest dimension. Control buttons shall be raised, flush, or recessed. Control buttons shall be arranged with numbers in ascending order. When two or more columns of buttons are provided they shall read from left to right. See Fig.	(1) Buttons. All control buttons shall be at least 3/4 in (19 mm) in their smallest dimension. They may be raised, flush or recessed. Buttons shall be arranged with numbers in ascending order as shown in Fig. 23(a) and shall read from left to right	(1) Buttons. All control buttons shall be at least 3/4 in (19 mm) in their smallest dimension. They may be raised or flush.

VERTICAL TRANSPORTATION ACCESSIBILITY STANDARDS
Comparison Chart

ADAAG	ICC/ANSI A117.1 1998	CABO/ANSI A117.1-1992	ANSI A117.1-1986	UFAS
(2) Tactile, Braille, and Visual Control Indicators. All control buttons shall be designated by Braille and by raised standard alphabet characters for letters, Arabic characters for numerals, or standard symbols as shown in Fig. 23(a), and as required in ASME A17.1-1990. Raised and Braille characters and symbols shall comply with 4.30. The call button for the main entry floor shall be designated by a raised star at the left of the floor designation (see Fig. 23(a)). All raised designations for control buttons shall be placed immediately to the left of the button to which they apply. Applied plates, permanently attached, are an acceptable means to provide raised control designations. Floor buttons shall be provided with visual indicators to show when each call is registered. The visual indicators shall be extinguished when each call is answered.	**407.2.11.2 Button Designations.** Except where provided in a standard telephone keypad arrangement, control buttons shall be identified by tactile characters complying with Section 703.2. Tactile characters and Braille shall be placed immediately to the left of the button to which they apply. The control button for the main entry floor, and control buttons other than remaining buttons with floor designations, shall be identified with tactile symbols complying with Table 407.2.11.2. Buttons with floor designations shall be provided with visible indicators to show that a call has been registered. The visible indication shall extinguish when the car arrives at the designated floor. Telephone-style keypads shall be in a standard telephone keypad arrangement, and shall be identified by characters complying with Section 703.4. The number five key shall have a single raised dot. The dot shall be 0.118 inch (3 mm) to 0.120 inch (3.05 mm) base diameter and in other aspects comply with Table 703.5. Characters shall be centered on the corresponding keypad button. A display shall be provided in the car with visible indicators to show registered car destinations. The visible indication shall extinguish when the car arrives at the designated floor. A standard five-pointed star shall be used to indicate the main entry floor.	B4.10.1.12(a). 4.10.1.12.2 Designations for control buttons shall comply with 4.28.2, 4.28.5 and 4.28.6. The call button for the main entry floor shall be designated by a star. Raised and Braille designations for control buttons shall be placed immediately to the left of the button to which the designations apply. See Fig. B4.10.1.12(c). Floor buttons shall be provided with visible indicators to show that a call has been registered. The visible indication shall cease when the call has been answered.	(2) Tactile and Visual Control Indicators. All control buttons shall be designated by raised standard alphabet characters for letters, Arabic characters for numerals, or standard symbols as shown in Fig. 23(a), and as required in ANSI/ASME A17.1-1984 and A17.1a-1985. Raised characters and symbols shall comply with 4.28. The call button for the main entry floor shall be designated by a raised star at the left of the floor designation (see Fig. 23(a)). All raised designations for control buttons shall be placed immediately to the left of the button to which they apply. Applied plates, permanently attached, are an acceptable means to provide raised control designations. Floor buttons shall be provided with visual indicators to show when each call is registered. The visual indicators shall be extinguished when each call is answered.	(2) Tactile and Visual Control Indicators. All control buttons shall be designated by raised standard alphabet characters for letters, Arabic characters for numerals, or standard symbols as shown in Fig. 23(a), and as required in ANSI A17.1-1978 and A17.1a-1979. Raised characters and symbols shall comply with 4.30. The call button for the main entry floor shall be designated by a raised star at the left of the floor designation (see Fig. 23(a)). All raised designations for control buttons shall be placed immediately to the left of the button to which they apply. Applied plates, permanently attached, are an acceptable means to provide raised control designations. Floor buttons shall be provided with visual indicators to show when each call is registered. The visual indicators shall be extinguished when each call is answered.
(3) Height. All floor buttons shall be no higher than 54 in (1370 mm) above	**407.2.11.3 Height.** Buttons with floor designations shall be 48 inches (1220 mm)	4.10.1.12.3 Floor buttons shall be located 54 in (1370 mm) maximum above the	(3) Height. All floor buttons shall be no higher than 54 in (1370 mm) above	(3) Height. All floor buttons shall be no higher than 48 in (1220 mm), unless

VERTICAL TRANSPORTATION ACCESSIBILITY STANDARDS
Comparison Chart

ADAAG	ICC/ANSI A117.1 1998	CABO/ANSI A117.1-1992	ANSI A117.1-1986	UFAS
the finish floor for side approach and 48 in (1220 mm) for front approach. Emergency controls, including the emergency alarm and emergency stop, shall be grouped at the bottom of the panel and shall have their centerlines no less than 35 in (890 mm) above the finish floor (see Fig. 23(a) and (b)).	maximum above the floor or ground. Emergency controls, including the emergency alarm, shall be grouped at the bottom of the panel. Emergency control buttons shall have their centerlines 35 inches (890 mm) minimum above the floor or ground. **EXCEPTION:** Where the elevator serves more than 16 openings and parallel approach is provided, buttons with floor designations shall be 54 inches (1370 mm) maximum above the floor or ground.	floor for parallel approach and 48 in (1220 mm) maximum for front approach. Emergency controls, including the emergency alarm, shall be grouped at the bottom of the panel. Emergency control buttons shall have their centerlines 35 in (890 mm) minimum above the floor. See Fig. B4.10.1.12(c).	the floor for side approach and 48 in (1220 mm) for front approach. Emergency controls, including the emergency alarm and emergency stop, shall be grouped at the bottom of the panel and shall have their centerlines no less than 35 in (890 mm) above the finish floor (see Fig. 23(a) and (b)).	there is a substantial increase in cost, in which case the maximum mounting height may be increased to 54 in (1370 mm), above the floor. Emergency controls, including the emergency alarm and emergency stop, shall be grouped at the bottom of the panel and shall have their centerlines no less than 35 in (890 mm) above the finish floor (see Fig. 23(a) and (b)).
(4) Location. Controls shall be located on a front wall if cars have center opening doors, and at the side wall or at the front wall next to the door if cars have side opening doors (see Fig. 23(c) and (d)).	**407.2.11.4 Clear Floor or Ground Space.** A clear floor or ground space complying with Section 305 shall be provided at controls.	4.10.1.12.4 Controls shall be located on a front wall if cars have center opening doors, and at the side wall or at the front wall next to the door if cars have side opening doors.	(4) Location. Controls shall be located on a front wall if cars have center opening doors, and at the side wall or at the front wall next to the door if cars have side opening doors (see Fig. 23(c) and (d)).	(4) Location. Controls shall be located on a front wall if cars have center opening doors, and at the side wall or at the front wall next to the door if cars have side opening doors (see Fig. 23(c) and (d)).
4.10.13* Car Position Indicators.	**407.2.12 Car Position Indicators.** In elevator cars, both audible and visible indicators shall be provided to identify the floor location of the car.	**4.10.1.13* Car Position Indicators.** In elevator cars, both audible and visible car floor location indicators shall be provided.	**4.10.13* Car Position Indicators.**	**4.10.13* Car Position Indicators.**
In elevator cars, a visual car position indicator shall be provided above the car control panel or over the door to show the position of the elevator in the hoistway. As the car passes or stops at a floor served by the elevators, the corresponding numerals shall illuminate, and an audible signal shall sound. Numerals shall be a minimum of 1/2 in (13 mm) high.	**407.2.12.1 Visible Indicators.** Indicator shall be above the car control panel or above the door. Numerals shall be 1/2 inch (13 mm) high minimum. As the car passes or stops at a floor served by the elevator, the corresponding character shall illuminate.	4.10.1.13.1 Visible. Indicator shall be located above the car control panel or above the door. Numerals shall be 1/2 (13 mm) minimum. As the car passes or stops at a floor served by the elevator, the corresponding character shall illuminate.	In elevator cars, a visual car position indicator shall be provided above the car control panel or over the door to show the position of the elevator in the hoistway. As the car passes or stops at a floor served by the elevators, the corresponding numerals shall illuminate and an audible signal shall sound. Numerals shall be a minimum of 1/2 in (13 mm) high.	In elevator cars, a visual car position indicator shall be provided above the car control panel or over the door to show the position of the elevator in the hoistway. As the car passes or stops at a floor served by the elevators, the corresponding numerals shall illuminate and an audible signal shall sound. Numerals shall be a minimum of 1/2 in (13 mm) high.
The audible signal shall be no less than 20 decibels with a frequency no higher than 1500 Hz. An automatic verbal announcement of the floor number at which a car stops or which a car passes may be substituted for the audible signal.	**407.2.12.2 Audible Indicators.** The audible signal shall be 10 dBA minimum above ambient, but shall not exceed 80 dBA maximum, measured at the annunciator. The signal shall be an automatic verbal announcement which announces the floor at which the car has stopped.	4.10.1.13.2 Audible. Indicator shall be 20 decibels minimum with a frequency of 1500 Hz maximum above ambient. Indicator shall be either an audible signal which sounds when the car passes and when a car stops at a floor served by the elevator, or an automatic verbal announcement which	The audible signal shall be no less than 20 decibels with a frequency no higher than 1500 Hz. An automatic verbal announcement of the floor number at which a car stops or at which a car passes may be substituted for the audible signal.	The audible signal shall be no less than 20 decibels with a frequency no higher than 1500 Hz. An automatic verbal announcement of the floor number at which a car stops or which a car passes may be substituted for the audible signal.

VERTICAL TRANSPORTATION ACCESSIBILITY STANDARDS
Comparison Chart

ADAAG	ICC/ANSI A117.1 1998	CABO/ANSI A117.1-1992	ANSI A117.1-1986	UFAS
	EXCEPTION: For elevators that have a rated speed of 200 fpm (1 m/s) or less, an audible signal with a frequency of 1500 Hz maximum which sounds as the car passes or stops at a floor served by the elevator shall be permitted.	announces the floor at which the car has stopped.		
4.10.14* Emergency Communications. If provided, emergency two-way communication systems between the elevator and a point outside the hoistway shall comply with ASME A17.1-1990. The highest operable part of a two-way communication system shall be a maximum of 48 in (1220 mm) from the floor of the car. It shall be identified by a raised symbol and lettering complying with 4.30 and located adjacent to the device. If the system uses a handset then the length of the cord from the panel to the handset shall be at least 29 in (735 mm). If the system is located in a closed compartment the compartment door hardware shall conform to 4.27, Controls and Operating Mechanisms. The emergency intercommunication system shall not require voice communication.	**407.2.13 Emergency Communications.** Emergency two-way communication systems between the elevator car and a point outside the hoistway shall comply with ASME/ANSI A17.1. The highest operable part of a two-way communication system shall comply with Section 308.3. If the device is in a closed compartment, the compartment door hardware shall comply with Section 309. Tactile symbols and characters complying with Section 703.2 shall be provided adjacent to the device. If the system uses a handset, the cord from the panel to the handset shall be 29 inches (735 mm) long minimum. The car emergency signaling device shall not be limited to voice communication. If instructions for use are provided, essential information shall be presented in both tactile and visual form complying with Section 703.	**4.10.1.14* Emergency Communications.** If provided, car emergency signaling devices between the elevator and a point outside the hoistway shall comply with ASME/ANSI A17.1. The highest operable part of a two-way communication system shall be 54 in (1370 mm) maximum above the floor for parallel approach and 48 in (1220 mm) maximum above the floor for front approach. If the device is located in a closed compartment, the compartment door hardware shall comply with 4.25. The device shall be identified by raised symbols and lettering complying with 4.28 and located adjacent to the device. If the system uses a handset, the cord from the panel to the handset shall be 29 in (735 mm) long minimum. The car emergency signaling device shall not be limited to voice communication. If instructions for use are provided, essential information shall be presented in both tactile and visual form.	**4.10.14* Emergency Communications.** If provided, car emergency signaling devices between the elevator and a point outside the hoistway shall comply with ANSI/ASME A17.1-1984 and A17.1a-1985. The highest operable part of a two-way communication system shall be a maximum of 54 in (1370 mm) above the floor for side approach and 48 in (1220 mm) for front approach. If the system is located in a closed compartment, the compartment door hardware shall comply with 4.25. It shall be identified by raised symbol and lettering complying with 4.28 and located adjacent to the device. If the system uses a handset, then the length of the cord from the panel to the handset shall be at least 29 in (735 mm). The car emergency signaling device shall not be limited to voice communication. If instructions for use are provided, essential information shall be presented in both tactile and visual form.	**4.10.14* Emergency Communications.** If provided, emergency two-way communication systems between the elevator and a point outside the hoistway shall comply with ANSI A17.1-1978 and A17.1a-1979. The highest operable part of a two-way communication system shall be a maximum of 48 in (1220 mm) from the floor of the car. It shall be identified by a raised or recessed symbol and lettering complying with 4.30 and located adjacent to the device. If the system uses a handset, then the length of the cord from the panel to the handset shall be at least 29 in (735 mm). If the system is located in a closed compartment, the compartment door hardware shall comply with 4.27, Controls and Operating Mechanisms. The emergency intercommunication system shall not require voice communication.
	407.3 New Destination-Oriented Elevators. Destination-oriented elevators shall comply with Section 407.2.1, 407.2.4 through 407.2.10, and 407.2.13. Such elevators shall also comply with Section 407.3.1 through 407.3.5 and ASME/ANSI A17.1. They shall be passenger elevators. **407.3.1 Call Buttons.** Call			

VERTICAL TRANSPORTATION ACCESSIBILITY STANDARDS
Comparison Chart

ADAAG	ICC/ANSI A117.1 1998	CABO/ANSI A117.1-1992	ANSI A117.1-1986	UFAS
	buttons shall be 35 inches (890 mm) minimum and 48 inches (1220 mm) maximum above the floor or ground, measured to the centerline of the buttons. A clear floor or ground space complying with Section 305 shall be provided. Call buttons shall be 3/4 inch (19 mm) minimum in their smallest dimension. Buttons shall be raised or flush. Objects beneath hall call buttons shall protrude 1 inch (25 mm) maximum into the clear floor or ground space. Destination-oriented elevator systems shall have a keypad or other means for the entry of destination information. Keypads, if provided, shall be in a standard telephone keypad arrangement, and shall be identified by characters complying with Section 703.4. The number five key shall have a single raised dot. The dot shall be 0.118 inch (3 mm) to 0.120 inch (3.05 mm) base diameter, and in other aspects comply with Table 703.5. Destination-oriented elevator systems shall be provided with visual and audible signals which indicate which elevator car to enter. Characters shall be centered on the corresponding keypad button. A display shall be provided in the car with visible indicators to show registered car destinations. The visible indication shall extinguish when the car arrives at the designated floor. A standards five-pointed star shall be used to indicate the main entry floor. **407.3.2 Hall Signals.** A visible and audible signal shall be provided to indicate a car destination corresponding with Section 407.3.1. The audible tone and verbal announcement shall be the same as those given at the call button or call			

VERTICAL TRANSPORTATION ACCESSIBILITY STANDARDS
Comparison Chart

ADAAG	ICC/ANSI A117.1 1998	CABO/ANSI A117.1-1992	ANSI A117.1-1986	UFAS
	button keypad, if provided. Each elevator in a bank shall have audible and visible means for differentiation. **407.3.2.1 Visible Signals.** Visible signals shall comply with Sections 407.3.2.1.1 through 407.3.2.1.3. **407.3.2.1.1 Height.** Hall signal fixtures shall be 72 inches (1830 mm) minimum above the floor or ground, measured to the centerline of the fixture. **407.3.2.1.2 Size.** The visible signal elements shall be 2-1/2 inches (64 mm) minimum in their smallest dimension. **407.3.2.1.3 Visibility.** Signals shall be visible from the floor area adjacent to the hoistway entrance. **407.3.3 Car Controls.** Emergency controls, including the emergency alarm, shall have their centerlines 35 inches (890 mm) minimum and 48 inches (1220 mm) maximum above the floor or ground. Buttons shall be 3/4 inch (19 mm) minimum in their smallest dimension. Buttons shall be raised or flush. Controls shall accommodate a forward reach or side reach complying with Section 308. **407.3.4 Car Position Indicators.** In elevator cars, audible and visible car location indicators shall be provided. **407.3.4.1 Visible Indicators.** Indicators shall be above the car control panel or above the door. Numerals shall be 1/2 inch (13 mm) high minimum. The visible indicators shall extinguish when the car arrives at the designated floor. **407.3.4.2 Audible Indicators.** An automatic			

VERTICAL TRANSPORTATION ACCESSIBILITY STANDARDS
Comparison Chart

ADAAG	ICC/ANSI A117.1 1998	CABO/ANSI A117.1-1992	ANSI A117.1-1986	UFAS
	verbal announcement which announces the floor at which the car has stopped shall be provided. The announcement shall be 10 dBA minimum above ambient and 80 dBA maximum, measured at the annunciator. **407.3.5 Elevator Car Identification.** In addition to the tactile signs required by Section 407.2.4, a tactile elevator car identification shall be placed immediately below the hoistway entrance floor designation. The characters shall be 2 inches (51 mm) high and shall comply with Section 703.2. **407.3.6 Door and Signal Timing for Hall Calls.** The minimum acceptable time from notification of the car assigned at the keypad until the door starts to close shall be calculated by the following equation, but shall not be less than 5 seconds: T = D/1.5 ft/s (D/455 mm/s) Where T = total time in seconds and D = distance in feet (millimeters) from the keypad to the centerline of the assigned hoistway door. **407.4 Limited-Use/Limited-Application Elevators.** Limited-use/limited-application elevators shall comply with Sections 407.4.1 through 407.4.10 and ASME/ANSI A17.1, Part XXV. **407.4.1 Automatic Operation.** Elevator operation shall be automatic. Each car shall automatically stop at a floor landing within a tolerance of 1/2 inch (13 mm) under rated loading to zero loading conditions. **407.4.2 Call Buttons.** Call buttons in elevator lobbies and halls shall be 35 inches (890 mm) minimum and 48			

VERTICAL TRANSPORTATION ACCESSIBILITY STANDARDS
Comparison Chart

ADAAG	ICC/ANSI A117.1 1998	CABO/ANSI A117.1-1992	ANSI A117.1-1986	UFAS
	inches (1220 mm) maximum above the floor or ground, measured to the centerline of the buttons. Such call buttons shall have visual signals to indicate when each call is registered and when each call is answered. Call buttons shall be 3/4 inch (19 mm) minimum in their smallest dimension, and shall be raised or flush. The button that designates the up direction shall be above the button that designates the down direction. Objects beneath hall call buttons shall protrude 1 inch (25 mm) maximum. **407.4.3 Hall Signals.** A visible and audible signal complying with Section 407.2.3 shall be provided in the car or at the hoistway entrance to indicate the direction of travel. **407.4.4 Tactile Characters on Hoistway Entrances.** Tactile character and Braille floor designations shall be provided on both jambs of elevator hoistway entrances and shall be 60 inches (1525 mm) above the floor or ground, measured from the baseline of the characters. A tactile star shall also be provided on both jambs at the main entry level. Such characters shall be 2 inches (51 mm) high and shall comply with Section 703.2. **407.4.5 Doors** Elevator hoistway doors shall be either swinging or horizontally sliding type. Elevator doors shall open and close automatically. Horizontally sliding type hoistway and car doors shall comply with Section 407.2.5. Swinging hoistway and car doors shall comply with Section 404. Swinging doors shall be low energy power-operated and shall comply with ANSI/BHMA A156.19. Power operated swinging			

VERTICAL TRANSPORTATION ACCESSIBILITY STANDARDS
Comparison Chart

ADAAG	ICC/ANSI A117.1 1998	CABO/ANSI A117.1-1992	ANSI A117.1-1986	UFAS
	doors shall remain open for 20 seconds minimum when activated.			
	407.4.6 Inside Dimensions of Elevator Cars. Elevator cars shall provide a clear width of 42 inches (1065 mm) minimum and a clear depth of 54 inches (1370 mm) minimum. Car doors shall be positioned at the narrow end(s) of the car and shall provide a clear width of 32 inches (815 mm) minimum.

EXCEPTION: For installations in existing buildings, elevator cars shall provide a clear width of 36 inches (915 mm) minimum, a clear depth of 54 inches (1370 mm) minimum, and a net clear platform area of 15 square feet (1.5 m²) minimum. | | | |
	407.4.7 Floor or Ground Surfaces. Floor or ground surfaces in elevator cars shall comply with Section 302. The horizontal distance between the car platform sill and the edge of any hoistway landing shall be 1-1/4 inches (32 mm) maximum.			
	407.4.8 Illumination Levels. The level of illumination at the car controls, platform, and car threshold and landing sill shall be 5 footcandles (54 lux) minimum.			
	407.4.9 Car Controls. Elevator car controls shall comply with Section 407.4.9.1 through 407.4.9.3.			
	407.4.9.1 Buttons. Control buttons shall be 3/4 inch (19 mm) minimum in their smallest dimension. Control buttons shall be raised or flush. Control buttons shall be arranged with numbers in ascending order.			
	407.4.9.2 Identification. Control buttons shall be identified by tactile characters complying with			

VERTICAL TRANSPORTATION ACCESSIBILITY STANDARDS
Comparison Chart

ADAAG	ICC/ANSI A117.1 1998	CABO/ANSI A117.1-1992	ANSI A117.1-1986	UFAS
	Section 703.2. Tactile characters shall be placed immediately to the left of the button to which they apply. The control button for the main entry floor shall be identified with a tactile symbol complying with Table 407.2.11.2. Buttons with floor designations shall be provided with visible indicators to show that a call has been registered. The visible indication shall extinguish when the car arrives at the designated floor. **407.4.9.3 Height.** Buttons with floor designations shall be 48 inches (1220 mm) maximum above the floor. Emergency controls, including the emergency alarm, shall be grouped at the bottom of the panel. Emergency control buttons shall have their centerlines 35 inches (890 mm) minimum above the floor. **407.4.9.4 Location.** Controls shall be on a side wall and a clear floor or ground space complying with Section 309.2 shall be provided. **407.4.10 Emergency Communications.** Emergency two-way communication systems between the elevator car and a point outside the hoistway shall comply with ASME/ANSI A17.1. The highest operable part of a two-way communication system shall comply with Section 308.3. If the device is in a closed compartment, the compartment door hardware shall comply with Section 309. Tactile symbols and characters complying with Section 703.2 shall be provided adjacent to the device. If the system uses a handset, the cord from the panel to the handset shall be 29 inches (735 mm) long			

VERTICAL TRANSPORTATION ACCESSIBILITY STANDARDS
Comparison Chart

ADAAG	ICC/ANSI A117.1 1998	CABO/ANSI A117.1-1992	ANSI A117.1-1986	UFAS
	minimum. The car emergency signaling device shall not be limited to voice communication. If instructions for use are provided, essential information shall be presented in both tactile and visual form complying with Section 703.			
	407.5 Existing Elevators. Accessible elements of existing elevators shall comply with Sections 407.5, 407.2.4, 407.2.6, 407.2.7, 407.2.9, 407.2.10, and 407.2.13. They shall be passenger elevators as classified by ASME/ANSI A17.1. **EXCEPTION:** Destination-oriented elevators which comply with Section 407.3.	**4.10.2 Elevators - Existing** **4.10.2.1 General.** Existing passenger elevators that are required to be accessible shall comply with 4.10.2 and with 4.10.1.2, 4.10.1.5, 4.10.1.7 through 4.10.1.11, and 4.10.1.14. All elevators that are programmed to respond to the same hall call control as the required accessible elevator shall comply with the requirements of 4.10.2.		
	407.5.1 Automatic Operation. Elevator operation shall be automatic. Each car shall be equipped with a self-leveling feature that will automatically bring and maintain the car at floor landings within a tolerance of 1/2 inch (13 mm) under rated loading to zero loading conditions.			
	407.5.2 Call Buttons. Call buttons in elevator lobbies shall 35 inches (890 mm) and 48 inches (1220 mm) maximum above the floor or ground, measured to the centerline of the button, where the appropriate floor or ground area complying with Section 305 is provided. The button that designates the up direction shall be above the button that designates the down direction. Keypad controls complying with Section 407.2.2 shall be permitted.	**4.10.2.2 Call Buttons.** The top of the hall call buttons shall be located vertically between 35 in (890 mm) and 54 in (1370 mm) above the floor when the appropriate floor area specified in 4.2.5 or 4.2.6 is provided. The button that designates the up direction shall be located above the button that designates the down direction.		
	407.5.3 Hall Signals. A visible and audible signal shall be provided at each hoistway entrance to indicate which car is answering a call,	**4.10.2.3 Hall Signals.** A visible and audible signal shall be provided at each hoistway entrance to indicate which car is answering a call,		

VERTICAL TRANSPORTATION ACCESSIBILITY STANDARDS
Comparison Chart

ADAAG	ICC/ANSI A117.1 1998	CABO/ANSI A117.1-1992	ANSI A117.1-1986	UFAS
	except that in-car signals complying with Section 407.2.3 shall be permitted. Audible signals shall sound once for the up direction and twice for the down direction, or shall have verbal annunciators that state the word "up" or "down." If new hall signals are provided, they shall comply with Section 407.2.3.	except that in-car signals complying with 4.10.1.4 shall be acceptable. Audible signals shall sound once for the up direction and twice for the down direction, or shall have verbal annunciators that state the word "up" or "down." If hall signals are added, they shall comply with 4.10.1.4.		
	407.5.4 Doors. Doors shall comply with Section 407.5.4.1 or 407.5.4.2.	**4.10.2.4 Door Operation.** Power operated horizontally sliding car and hoistway doors opened and closed by automatic means shall comply with 4.10.1.6. Existing manually operated hoistway swing doors shall comply with 4.13.5 and 4.13.11. A power operated car door that opens and maintains a 32 in (815 mm) minimum clear width shall be provided. Closing of the car door shall not be initiated until the hoistway door is closed. Car gates are prohibited.		
	407.5.4.1 Power Operated Doors. Power operated horizontally sliding car and hoistway doors opened and closed by automatic means shall comply with Section 407.2.5.			
	407.5.4.2 Manually Operated Doors. Existing manually operated hoistway swinging doors shall comply with Sections 404.2.3 and 404.2.9. A power operated car door that opens and maintains a 32 inches (815 mm) minimum clear width shall be provided. Closing of the car door shall not be initiated until the hoistway door is closed. Car gates are prohibited.			
	407.5.5 Inside Dimension of Elevator Cars. The inside dimension of elevator cars shall comply with Section 407.2.8. **EXCEPTION:** Existing car			

VERTICAL TRANSPORTATION ACCESSIBILITY STANDARDS
Comparison Chart

ADAAG	ICC/ANSI A117.1 1998	CABO/ANSI A117.1-1992	ANSI A117.1-1986	UFAS
	configurations that provide a clear floor area of 16 square feet (1.5 m²) minimum, and provide 48 inches (1220 mm) minimum inside clear depth and a 36 inches (915 mm) minimum clear width.			
	407.5.6 Car Controls. Elevator controls shall comply with Sections 407.5.6.1 through 407.5.6.4.	**4.10.2.5 Car Controls.** Elevator control panels shall have the following features:		
	407.5.6.1 Buttons. Car control buttons shall be 3/4 inch (19 mm) minimum in their smallest dimension. Control buttons shall be raised, flush or recessed. Where the car operating panel is changed, control buttons shall comply with Section 407.2.11.1.	**4.10.2.5.1** Car control buttons shall be 3/4 in (19 mm) minimum in their smallest dimension. Control buttons shall be raised, flush or recessed.		
		4.10.2.5.2 When the car operating panel is changed, control buttons shall comply with 4.10.1.12.1.		
	407.5.6.2 Designations and Indicators for Control Buttons. All control buttons shall comply with Section 407.2.11.2. EXCEPTION: Where existing car operating panel construction precludes locating tactile markings to the left of the controls, markings shall be placed as near to the control as possible.	**4.10.2.5.3** All control buttons shall comply with 4.10.1.12.2. EXCEPTION: When existing car operating panel construction precludes locating tactile markings to the left of the controls, markings shall be placed as near to the control as possible.		
	407.5.6.3 Height. All buttons with floor designations shall be 54 inches (1370 mm) maximum above the floor for parallel approach and 48 inches (1220 mm) maximum above the floor for forward approach. When the panel is changed, it shall comply with Section 407.2.11.3.	**4.10.2.5.4** All floor buttons shall be located 54 in (1370 mm) maximum above the floor for parallel approach and 48 in (1220 mm) maximum above the floor for front approach. When the panel is changed, emergency controls, including the emergency alarm, shall comply with 4.10.1.12.3.		
		4.10.2.5.5 Location of controls shall comply with 4.10.1.12.4.		
	407.5.6.4 Operating Panels. Where a new car operating panel complying	**4.10.2.5.6** When a new car operating panel conforming to the requirements of		

VERTICAL TRANSPORTATION ACCESSIBILITY STANDARDS
Comparison Chart

ADAAG	ICC/ANSI A117.1 1998	CABO/ANSI A117.1-1992	ANSI A117.1-1986	UFAS
	with the requirements of Section 407.2.11 is provided, existing car operating panels not complying with Section 407.2.11 are not required to be removed.	4.10.1.12 is provided, existing car operating panel(s) not conforming to 4.10.1.12 are not required to be removed.		
	407.5.7 Car Position Indicators. Where a new car position indicator is provided, the indicator shall comply with Section 407.2.12.	**4.10.2.6 Car Position Indicators.** When a new car position indicator is installed, the indicator shall comply with 4.10.1.13.		
	407.5.8 Identification. Elevators that comply with Section 407.5 shall be clearly identified with the International Symbol of Accessibility complying with Section 703.7, unless all elevators in the building are accessible.	**4.10.2.7 Identification.** Elevators that comply with this standard shall be clearly identified with the international symbol of accessibility, unless all elevators in the building are accessible. See Fig. 4.28.8.1.		
4.11 Platform Lifts (Wheelchair Lifts). **4.11.1 Location.** Platform lifts (wheelchair lifts) permitted by 4.1 shall comply with the requirements of 4.11.	**408 Wheelchair (Platform) Lifts**	**4.11 Wheelchair Lifts**	**4.11 Platform Lifts** **4.11.2 Requirements.** Platform lifts on an accessible route shall comply with 4.2.4, 4.5, and 4.25.	**4.11* Platform Lifts.** **4.11.1 Location.** Platform lifts permitted by 4.1 shall comply with the requirements of 4.11.
4.11.2* Other Requirements. If platform lifts (wheelchair lifts) are used, they shall comply with 4.2.4, 4.5, 4.27 and ASME A17.1 Safety Code for Elevators and Escalators, Section XX, 1990.	**408.1 General.** Wheelchair (Platform) lifts shall comply with ASME/ANSI A17.1 and with Sections 302, 305 and 309. Wheelchair (platform) lifts shall not be attended-operated and shall provide unassisted entry and exit from the lift.	Wheelchair lifts, if provided, shall comply with ASME/ANSI A17.1 and with 4.2.4, 4.5, and 4.25. Wheelchair lifts shall not require an attendant for operation.	**4.11.1 General.** Platform lifts complying with ANSI/ASME A17.1-1984 and A17.1a-1985 or the applicable safety regulations of administrative authorities having jurisdiction may be used as part of an accessible route.	**4.11.2 Other Requirements.** If platform lifts are used, they shall comply with 4.2.4, 4.5, 4.27, and the applicable safety regulations of administrative authorities having jurisdiction.
4.11.3 Entrance. If platform lifts are used then they shall facilitate unassisted entry, operation, and exit from the lift in compliance with 4.11.2.	**408.2 Doors and Gates.** Lifts shall have low energy power-operated doors or gates complying with Section 404.3. Doors and gates shall remain open for 20 seconds minimum. End doors shall be 32 inches (815 mm) minimum clear width. Side doors shall be 42 inches (1065 mm) minimum clear width. **EXCEPTION:** Lifts having doors or gates on opposite sides shall be permitted to have manual doors or gates. **1002 Type A Dwelling Units.** **1002.1 General** Type A dwelling units shall comply			**4.11.3 Entrance.** If platform lifts are used, then they should facilitate unassisted entry and exit from the lift in compliance with 4.11.2.

VERTICAL TRANSPORTATION ACCESSIBILITY STANDARDS
Comparison Chart

ADAAG	ICC/ANSI A117.1 1998	CABO/ANSI A117.1-1992	ANSI A117.1-1986	UFAS
	with Section 1002. **1002.7 Private Residence Elevators.** Elevators shall comply with Sections 1002.7.1 through 1002.7.8. **EXCEPTION:** Elevators complying with Section 407. **1002.7.1 Automatic Operation.** Elevator operation shall be automatic. Each car shall automatically stop at a floor landing within a tolerance of 1/2 inch (13 mm) under rated loading to zero loading conditions. **1002.7.2 Call Buttons.** Call buttons at elevator landings shall comply with Section 309.3. Call buttons shall be 3/4 inch (19 mm) minimum in their smallest dimension. **1002.7.3 Doors and Gates.** Elevator car and hoistway doors and gates shall comply with Section 404, except that the maneuvering clearances required by Section 404.2.4.1 for approaches to the push side of swinging doors shall not apply. Elevator car doors and gates shall be power operated. For elevators with a car that has more than one opening, the hoistway doors and gates shall be permitted to be of the manual-open, self-close type. Elevators with a single opening car shall have low-energy power-operated hoistway doors and gates. Power operated doors and gates shall comply with ANSI/BHMA A156.19 and shall remain open for 20 seconds minimum when activated. **1002.7.4 Inside Dimensions of Elevator Cars.** Elevator cars shall provide a clear floor space of 30 inches (760 mm) minimum by 48 inches (1220 mm) minimum. Car gates or doors shall be positioned at the narrow end of the clear floor space. The clearance			

VERTICAL TRANSPORTATION ACCESSIBILITY STANDARDS
Comparison Chart

ADAAG	ICC/ANSI A117.1 1998	CABO/ANSI A117.1-1992	ANSI A117.1-1986	UFAS
	between the car platform sill and the edge of any hoistway landing shall be 1-1/4 inches (32 mm) maximum. **1002.7.5 Floor Surfaces.** Floor surfaces in elevator cars shall comply with Sections 302 and 303. **1002.7.6 Illumination Levels.** The level of illumination at the car controls, platform, and car threshold and landing sill shall be 5 footcandles (54 lux) minimum. **1002.7.7 Car Controls.** Elevator car controls shall comply with Sections 1002.7.7.1 through 1002.7.7.3. **1002.7.7.1 Buttons.** Control buttons shall be 3/4 inch (19 mm) minimum in their smallest dimension. Control buttons shall be raised or flush. **1002.7.7.2 Height.** Buttons with floor designations shall comply with Section 309.3. **1002.7.7.3 Location.** Controls shall be on a side wall, 12 inches (305 mm) minimum from any adjacent wall. **1002.7.8 Emergency Communications.** A telephone and emergency signal device shall be provided in the car and shall comply with ASME/ANSI A17.1, Rule 509. The telephone and emergency signaling device shall comply with Section 309.3. If the device is in a closed compartment, the compartment door hardware shall comply with Section 309. The telephone cord shall be 29 inches (735 mm) long minimum. **1003 Type B Dwelling Units.** **1003.1 General.** Type B dwelling units shall comply with Section 1003.			

VERTICAL TRANSPORTATION ACCESSIBILITY STANDARDS
Comparison Chart

ADAAG	ICC/ANSI A117.1 1998	CABO/ANSI A117.1-1992	ANSI A117.1-1986	UFAS
	1003.7 Private Residence Elevators. Elevators shall comply with Section 1002.7. **EXCEPTION:** Elevators complying with Section 407.			
10. TRANSPORTATION FACILITIES. **10.1 General.** Every station, bus stop, bus stop pad, terminal, building or other transportation facility, shall comply with the applicable provisions of 4.1 through 4.35, sections 5 through 9, and the applicable provisions of this section. The exception for elevators in 4.1.3(5), exception 1 and 4.1.6(1)(k) do not apply to a terminal, depot, or other station used for specified public transportation, or an airport passenger terminal, or facilities subject to Title II. **10.3 Fixed Facilities and Stations.** **10.3.1 New Construction.** New stations in rapid rail, light rail, commuter rail, intercity bus, intercity rail, high speed rail, and other fixed guideway systems (e.g., automated guideway transit, monorails, etc.) shall comply with the following provisions, as applicable: (16) Where provided in below grade stations, escalators shall have a minimum clear width of 32 inches. At the top and bottom of each escalator run, at least two contiguous treads shall be level beyond the comb plate before the riser begin to form. All escalator treads shall be marked by a strip of clearly contrasting color, 2 inches in width, placed parallel to and on the nose of each step. The strip shall be of a material that is at least as slip resistant as the remainder of the tread. The edge of the tread shall be apparent from both				

VERTICAL TRANSPORTATION ACCESSIBILITY STANDARDS
Comparison Chart

ADAAG	ICC/ANSI A117.1 1998	CABO/ANSI A117.1-1992	ANSI A117.1-1986	UFAS
ascending and descending directions. (17) Where provided, elevators shall be glazed or have transparent panels to allow an unobstructed view both in to and out of the car. Elevators shall comply with 4.10. EXCEPTION: Elevator cars with a clear floor area in which a 60 inch diameter circle can be inscribed may be substituted for the minimum car dimensions of 4.10, Fig. 22.				
APPENDIX A **A4.10 Elevators.**		**APPENDIX A** **A4.10 Elevators** **A4.10.1.5 Raised Characters on Hoistway Entrances.** Plates that have the appropriate raised characters are acceptable provided the plates are permanently fixed to the hoistway entrance jambs.	**APPENDIX A** **A4.10 Elevators**	**APPENDIX A** **A4.10 Elevators**
A4.10.6 Door Protective and Reopening Device. The required door reopening device would hold the door open for 20 seconds if the doorway remains obstructed. After 20 seconds, the door may begin to close. However, if designed in accordance with ASME A17.1-1990, the door closing movement could still be stopped if a person or object exerts sufficient force at any point on the door edge.		**A4.10.1.6 Door Protective and Reopening Device.** The required door reopening device holds the door open for 20 seconds if the doorway remains obstructed. After 20 seconds, the door begins to close. However, if designed in accordance with ASME/ANSI A17.1, the door closing movement is still stopped if a person or object exerts sufficient force at any point on the door edge. Owing to the kinetic nature of the motion, reversal of the closing door is not instantaneous. Until the continued movement of the door is arrested, it is possible that limited movement of the door causes it to come in contact with a person or object in its path.	**A4.10.6 Door Protective and Reopening Device.** The required door reopening device would hold the door open for 20 seconds if the doorway remains obstructed. After 20 seconds, the door may begin to close. However, if designed in accordance with ANSI/ASME A17.1-1984 and A17.1a-1985, the door closing movement could still be stopped if a person or object exerts sufficient force at any point on the door edge. Owing to the kinetic nature of the motion, reversal of the closing door is not instantaneous. Until the continued movement of the door is arrested, it is possible that limited movement of the door may cause it to come in contact with a person or object in its path.	**A4.10.6 Door Protective and Reopening Device.** The required door reopening device would hold the door open for 20 seconds if the doorway remains unobstructed. After 20 seconds, the door may begin to close. However, if designed in accordance with ANSI A17.1-1978, the door closing movement could still be stopped if a person or object exerts sufficient force at any point on the door edge.
A4.10.7 Door and Signal Timing for Hall Calls. This paragraph allows variation in the location of call buttons, advance time for warning signals, and the door-holding		**A4.10.1.7 Door and Signal Timing for Hall Calls.** This subsection allows variation in the location of call buttons, advance time for warning signals, and the door-holding	**A4.10.7 Door and Signal Timing for Hall Calls.** This subsection allows variation in the location of call buttons, advance time for warning signals, and the door-holding	**A4.10.7 Door and Signal Timing for Hall Calls.** This paragraph allows variation in the location of call buttons, advance time for warning signals, and the door-holding

VERTICAL TRANSPORTATION ACCESSIBILITY STANDARDS
Comparison Chart

ADAAG	ICC/ANSI A117.1 1998	CABO/ANSI A117.1-1992	ANSI A117.1-1986	UFAS
period used to meet the time requirement.	period used to meet the time requirement. Examples of the application of this provision are shown in Fig. BA4.10.1.7. **A4.10.1.9 Inside Dimensions of Elevator Cars.** See Fig. BA4.10.1.9 for one example of acceptable elevator car dimensions. Elevator car floor plans that may meet the intent of the criteria are available from: National Elevator Industry, Inc., 185 Bridge Plaza North, Room 310, Fort Lee, New Jersey 07024.	period used to meet the time requirement.	period used to meet the time requirement.	period used to meet the time requirement.
A4.10.12 Car Controls. Industry-wide standardization of elevator control panel design would make all elevators significantly more convenient for use by people with severe visual impairments. In many cases, it will be possible to locate the highest control on elevator panels within 48 in (1220 mm) from the floor.	**A4.10.1.12 Car Controls.** Industry-wide standardization of elevator control panel design makes all elevators significantly more convenient for use by people with severe visual impairments. In many cases, it is possible to locate the highest control on elevator panels within 48 in (1220 mm) from the floor. Permanently applied plates that have the appropriate raised characters and symbols are an acceptable means of providing raised control designations.	**A4.10.12 Car Controls.** Industry-wide standardization of elevator control panel design would make all elevators significantly more convenient for use by people with severe visual impairments. In many cases, it will be possible to locate the highest control on elevator panel within 48 in (1220 mm) from the floor.	**A4.10.12 Car Controls.** Industry-wide standardization of elevator control panel design would make all elevators significantly more convenient for use by people with severe visual impairments. In many cases, it will be possible to locate the highest control on elevator panel within 48 in (1220 mm) from the floor.	**A4.10.12 Car Controls.** Industry-wide standardization of elevator control panel design would make all elevators significantly more convenient for use by people with severe visual impairments. In many cases, it will be possible to locate the highest control on elevator panel within 48 in (1220 mm) from the floor.
A4.10.13 Car Position Indicators. A special button may be provided that would activate the audible signal within the given elevator only for the desired trip, rather than maintaining the audible signal in constant operation.	**A4.10.1.13 Car Position Indicators.** A verbal announcement can serve as an audible signal and is preferred by persons with visual impairments and the general public. A non-verbal, audible signal is difficult to use in high rise buildings and when non-standard floor arrangements (i.e. basement, lobby, mezzanine, 2nd floor, ... 12th floor, 14th floor, etc.) are utilized. A special button is sometimes provided that activates the audible signal within the given elevator only for the desired trip, rather than maintaining the audible	**A4.10.13 Car Position Indicators.** A special button may be provided that would activate the audible signal within the given elevator only for the desired trip, rather than maintaining the audible signal in constant operation.	**A4.10.13 Car Position Indicators.** A special button may be provided that would activate the audible signal within the given elevator only for the desired trip, rather than maintaining the audible signal in constant operation.	**A4.10.13 Car Position Indicators.** A special button may be provided that would activate the audible signal within the given elevator only for the desired trip, rather than maintaining the audible signal in constant operation.

VERTICAL TRANSPORTATION ACCESSIBILITY STANDARDS
Comparison Chart

ADAAG	ICC/ANSI A117.1 1998	CABO/ANSI A117.1-1992	ANSI A117.1-1986	UFAS
	signal in constant operation. The elevator industry recommends the button be identified by the symbol "S" and be located with, immediately above or immediately below the emergency buttons on the car operating panel.			
A4.10.14 Emergency Communications. A device that requires no handset is easier to use by people who have difficulty reaching. Also, small handles on handset compartment doors are not usable by people who have difficulty grasping. Ideally, emergency two-way communication systems should provide both voice and visual display intercommunication so that persons with hearing impairments and persons with vision impairments can receive information regarding the status of a rescue. A voice intercommunication system cannot be the only means of communication because it is not accessible to people with speech and hearing impairments. While a voice intercommunication system is not required, at a minimum, the system should provide both an audio and visual indication that a rescue is on the way. **A4.11 Platform Lifts (Wheelchair Lifts).** **A4.11.2 Other Requirements.** Inclined	**A4.10.1.14 Emergency Communications.** A device that requires no handset is easier to use by people who have difficulty reaching.	**A4.10.14 Emergency Communications.** A device that requires no handset is easier to use by people who have difficulty reaching.	**A4.10.14 Emergency Communications.** A device that requires no handset is easier to use by people who have difficulty reaching.	**A4.11 Platform Lifts.** Platform lifts include porch lifts and other devices used for short-distance, vertical transportation of people in wheelchairs. At the present time, generally recognized safety standards for such lifts have not been developed. Care should be taken in selecting and installing lifts to ensure that they are free from hazards to users or to other individuals who may be in the vicinity where they are being operated.

VERTICAL TRANSPORTATION ACCESSIBILITY STANDARDS
Comparison Chart

ADAAG	ICC/ANSI A117.1 1998	CABO/ANSI A117.1-1992	ANSI A117.1-1986	UFAS
stairway chairlifts, and inclined and vertical platform lifts (wheelchair lifts) are available for short-distance, vertical transportation of people with disabilities. Care should be taken in selecting lifts as some lifts are not equally suitable for use by both wheelchair users and semi-ambulatory individuals.				

NOTES:
(1) ADAAG - Published in Federal Register July 26, 1991 and amended September 6, 1991.
(2) ICC/ANSI A117.1-1998 copyrighted by International Code Council.
(3) CABO/ANSI A117.1-1992 copyrighted Council of American Building Officials.
(4) ANSI A117.1-1986 copyrighted American National Standards Institute
(5) UFAS published by US Government Printing Office
(5) See referenced document for all tables and figures.

ADAAG ELEVATOR CHECKLIST

7.

ADAAG ELEVATOR CHECKLIST

LOCATION:_____

CITY:_____ STATE:_____ ELEVATOR NO.:_____

No.	ADAAG Section	Item	Findings and Notes
		GENERAL	
1*	4.10.1	Elevator must comply with ASME A17.1-1990. Freight elevator are not acceptable unless only elevator provided and is permitted to carry passengers both public and employees.	
		AUTOMATIC OPERATION	
2	4.10.2	Elevators must be Automatic.	
3	4.10.2	Self-leveling to within 1/2 in.	
		HALL CALL BUTTONS	
4	4.10.3	Buttons centered at 42 in. above the floor.	
5	4.10.3	Buttons to illuminate when call is entered and extinguish when answered.	
6	4.10.3	Buttons to be at least 3/4 in. in the smallest dimension.	
7	4.10.3	Up button located above down button.	
8	4.10.3	Buttons raised or flushed.	
9	4.10.3	Objects mounted beneath hall buttons not to project into the lobby more than 4 in.	
		HALL LANTERNS	
10	4.10.4	Visible and audible signals at each hoistway entrance to indicate which car is responding to the call.	
11	4.10.4	Audible signals to sound once for up and twice for "down" or may verbal announcement stating "up" "down."	
12	4.10.4(1)	Hall directional lantern centered 72 in. above floor.	

ADAAG ELEVATOR CHECKLIST

No.	ADAAG Section	Item	Findings and Notes
13	4.10.4(2)	Directional lantern visible elements minimum of 2-½ in. in the smallest dimension.	
14	4.10.4(3)	Directional lanterns must be visible from the vicinity of the hall call button.	
15	4.10.4(4)	In car lanterns, meeting the requirements above are acceptable in lieu of hall directional lanterns.	
		HOISTWAY ENTRANCES	
16	4.10.5	Raised and Braille floor designations are required on both door jambs. Permanently applied plates are acceptable.	
17	4.10.5	Centerline of floor designation characters 60 in. above floor.	
18	4.30.4	Characters must be 2 in. high, raised 1/32 in. upper sans serif (block letters) or simple serif type.	
19	4.30.4	Grade II Braille to accompany raised characters.	
		DOOR PROTECTIVE & REOPENING DEVICES	
20	4.10.6	Doors must open and close automatically.	
21	4.10.6	Non-contact door reopening device at 5 in. and 29 in. above the floor.	
22	4.1.6(3)(c)	If safety edges are provided on existing elevators, the non-contact door reopening devices may be omitted.	
23	4.10.6	Reopening device to remain operational for at least 20 seconds.	

ADAAG ELEVATOR CHECKLIST

No.	ADAAG Section	Item	Findings and Notes
		DOOR AND SIGNAL TIMING	
24	4.10.7	Minimum acceptable door open time from notification car is answering a hall call until the car doors begin to close: $$T=D/(1.5ft/s)$$ where T is the total time in and D is the distance from a point in the lobby or corridor 60 in. directly in front of the farthest button controlling that car to centerline of its hoistway door.	
25	4.10.7	Minimum acceptable notification time 5 seconds.	
		DOOR DELAY FOR CAR CALLS	
26	4.10.8	Doors to remain open for a minimum of seconds in response to car calls.	
		FLOOR PLAN NEW ELEVATOR	
27	4.10.9	Shall provide space for wheelchair users to enter the car, within reach of controls, and the car. Doors shall provide 36 in. minimum opening. Car depth: 51 in. minimum, with 54 in. minimum from rear of cab to inside face of door Car Width: Side-opening door 68 in. minimum; center-opening door 80 in. minimum.	

ADAAG ELEVATOR CHECKLIST

No.	ADAAG Section	Item	Findings and Notes
		FLOOR PLAN EXISTING ELEVATOR	
28	4.1.6(3)(c	Where existing configuration or technical infeasibility prevent use car size specified in check Item 27, dimensions may be reduced as required. However, in no case, shall inside car dimensions be less than 48 in. by 48 in. Equivalent facilitation may be provided with a cab of different size when usability can be demonstrated and when all other items required to be accessible comply.	
29	4.10.9	Clearance between car platform sill and edge of hoistway landing sill no greater than 1-¼ in.	
		FLOOR SURFACES	
30	4.10.10 & 4.5.1	Surfaces to be stable, firm and slip resistant.	
31	4.5.3	Carpeting if installed, must have firm cushion, pad or backing, or no cushion or pad.	
32	4.5.3	Carpeting must have level loop, textured loop, level pile texture.	
33	4.5.3	Carpeting pile thickness not to exceed 1/2 in.	
34	4.5.3	Carpeting must have exposed edges fastened to the floor surface.	
35	4.5.3	Exposed edges of carpets must be trimmed.	
		ILLUMINATION LEVELS	
36	4.10.11	Five footcandles of illumination to be provided at car controls, platform and at sill.	
		CAR CONTROLS	
37	4.10.12(1)	Buttons to be at least 3/4 in. in their smallest dimension.	

ADAAG ELEVATOR CHECKLIST

No.	ADAAG Section	Item	Findings and Notes
38	4.10.12(1)	Buttons must be flush or raised.	
39	4.10.12(2)	Buttons must be designated by raised characters and Braille or symbols complying with ASME A17.1 Rule 210.13.	
40	4.10.12(2)	Characters must be a minimum of 5/8 in. high, upper case sans (block letters) or simple serif type.	
41	4.10.12(2)	Grade II Braille to accompany raised character of symbol.	
42	4.10.12(2)	Raised designations must be to the immediate left of the button to which they apply.	
43	4.10.12(2)	Call button illuminates when call is entered and extinguish when answered.	
44	4.10.12(3)	Floor buttons must be no higher then 48 in. when located in front return. Buttons must be no higher than 54 in. when a side approach provided.	
45	4.10.12(3)	Emergency controls, including emergency alarm and emergency stop (if provided) must be grouped at the bottom of the panel and have centerlines no less than 35 in. above the finished floor.	
46	4.10.12(4)	Controls must be on the front return wall with center-opening doors. They may be on the front return or strike jamb side wall with side doors.	
		CAR POSITION INDICATORS	
47	4.10.13	Visual car position indicator must be provided above control panel or over door.	
48	4.10.13	Car position indicator numerals must be a minimum of 1/2 in. high.	

ADAAG ELEVATOR CHECKLIST

No.	ADAAG Section	Item	Findings and Notes
49	4.10.13	Audible signal to sound as the car passes or stops at a floor and a corresponding floor designation must illuminate. Audible signal must be at least 20 dB with a frequency no higher than 1,500 Hz.	
50	4.10.13	A button to activate audible signal only for desired trip may be provided.	
51	4.10.13	An automatic verbal announcement the floor at which a car stops may be substituted for the audible signal.	
		EMERGENCY COMMUNICATIONS	
52	4.10.14	If provided, emergency two-way communication systems between the elevator and a point outside the hoistway must comply with ASME A17.1-1990, Rule 211.1.	
53	4.10.14	The highest operable part must be a maximum of 48 in. from the car floor.	
54	4.10.14	Emergency communication identification must be provided and located adjacent to the device. Characters must be a minimum of 5/8 in. high raised 1/32 in., upper case serif (block letters) or simple serif type, and accompanied by Grade II Braille.	
55	4.10.13	If a handset is provided the cord must be at least 29 in. long.	
56	4.27.4	If located in a closed compartment, the door must be operable with one hand. It must not require tight grasping, pinching or twisting of the wrist. The force required to open the door must not exceed 5 lb/f.	
57	4.10.13	The system must not require voice communication.	

ADAAG ELEVATOR CHECKLIST

No.	ADAAG Section	Item	Findings and Notes
		TRANSPORTATION FACILITIES	
58	10.3.1	New stations in rapid rail, light rail, commuter rail, intercity bus, intercity rail, high speed rail and other fixed guideway systems (e.g. automated guideway transit, monorails etc.).	
		FLOOR PLAN OF ELEVATOR CARS	
59	10.3.1(17)	Elevator cars with a clear floor area in which a 60 in. diameter can be inscribed can be substituted for the dimensions specified in checklist Item 27.	
		GLAZING	
60	10.3.1(17)	Elevators to be glazed or have transparent panels to allow an unobstructed view into and out of the car.	
		ESCALATORS (Below Grade Stations)	
61	10.3.1(16)	Minimum clear width 32 in.	
62	10.3.1(16	Top and bottom of escalator to have at least two contiguous treads level beyond the combplate before risers begin to form	
63		Escalator treads must be marked by a strip of clear contrasting color, 2 in. wide, placed parallel to and on the nose of each step. The strip must be of a material that is at least as slip resistant as the remainder of the tread.	
64	10.3.1(16)	The edge of the tread must be apparent from both the ascending and descending directions.	

APPENDICES

APPENDIX A
REFERENCE MATERIAL

- Americans with Disabilities Act Accessibility Guidelines (ADAAG)
- UFAS Retrofit Manual
- ADA Accessibility Guidelines Checklist for Building and Facilities
- ADAAG Manual
    United States Architectural and Transportation Barriers Compliance Board
    131 F Street, NW, Suite 1000
    Washington, DC 20004-1111
    Telephone (202)272-5434
    www.access-board.gov
- ADA Regulations
- The Americans with Disabilities Act, Title II Technical Assistance Manual
- The Americans with Disabilities Act, Title III Technical Assistance Manual
    Office of the Americans with Disabilities Act
    Civil Right Division
    U.S. Department of Justice
    P. O. Box 66118
    Washington, DC 20035-6118
    Telephone (202) 514-0301
    www.usdoj.gov/crt/ada/adahom1.htm
- ADA Compliance Guide Book. A Checklist for Your Building
- The ADA Answer Book
    Building Owners and Managers Association (BOMA) International
    1201 New York Avenue, NW, Suite 300
    Washington, DC 20005
    Telephone (202) 408-2662
    www.boma.org
- Elevator Code Authorities Telephone Directory
    Edward A. Donoghue Associates Inc.
    Code and Safety Consultants
    1677 County Route 64
    P. O. Box 201
    Salem, NY 12865-0201
    Telephone (518) 854-9249
    www.eadai.com
- Fair Housing Accessibility Guidelines
    U. S. Department of Housing and Urban Development
    451 Seventh Street, SW
    Washington, DC 20410-0500

Telephone (202) 708-2618
www.hud.gov
- International Code Council (information on ICC/ANSI A117.1-1998 & International Build Code)
  5203 Leesburg Pike, Suite 708
  Falls Church, VA 22041
  Telephone (703) 931-4533
  www.intlcode.org
- National Building Code
- American National Standard for Accessible and Usable Building and Facilities, ICC/ANSI A117.1-1998 and CABO/ANSI A117.1-1992
  Building Officials and Code Administrators International (BOCA)
  4051 West Flossmoor Road
  Country Club Hills, IL 60478
  Telephone (708) 799-2300
  www.bocai.org
- Safety Code for Elevators and Escalators, ASME A17.1
- ASME A17.1 - Handbook, by Edward A. Donoghue, CPCA
- Safety Code for Existing Elevators and Escalators ASME A17.3
  American Society of Mechanical Engineers
  22 Law Drive
  P. O. Box 2300
  Fairfield, NJ 07007-2300
  Telephone 1-800-843-2763
  www.asme.org
- Standard Building Code
- American National Standard for Accessible and Usable Building and Facilities, ICC/ANSI A117.1-1998 and CABO/ANSI A117.1-1992
  Southern Building Code Congress International
  5200 Montclair Road
  Birmingham, IL 35213
  Telephone (205) 591-1853
  www.sbcci.org
- Uniform Building Code
- American National Standard for Accessible and Usable Building and Facilities, ICC/ANSI A117.1-1998 and CABO/ANSI A117.1-1992
  International Conference of Building Officials
  5360 South Workman Mill Road
  Whittier, CA 90601
  Telephone (213) 699-0541
  www.icbo.org
- Uniform Federal Accessibility Standards
  Superintendent of Documents

U.S. Government Printing Office
Washington, DC 20402
Telephone (202) 783-3238
www.gpo.gov
- Workshop on Elevators Use During Fires, NISTIR 4993
    U.S. Department of Commerce
    National Institute of Standards and Technology
    Gaithersburg, MD 20899
    www.nist.gov
- Vertical Transportation Standards
- Building Transportation Standards and Guidelines, NEII-1-2000 (available spring 2000)
    National Elevator Industry, Inc.
    Atrium at Glenpointe
    400 Frank W. Burr Boulevard
    Teaneck, NJ 07666-6801
    Telephone (201) 928-2828

APPENDIX B
SIGNAGE STANDARDS

B.1 HISTORY

The 1976 edition of NEII handicapped standard introduced requirements for door jamb and car control markings. Door jams markings were originally required to be 2½ in. high. The purpose of the door jamb marking is two-fold. The size of the marking permits those with a vision impairment to read the sign. Those that are blind can tactually read the markings. Further research determined that markings greater than 2 in. posed a problem for tactually reading. When tactually tracing, with your fingers, markings larger than 2 in., one cannot visualize the outline of the letter or numeral. Letters and numerals larger than 2 in. are unreadable by the blind.

The 1980 edition of ANSI A117.1 and all later editions, ADAAG and UFAS all specify that door jamb markings shall be a nominal 2 in. in height. All of the standards also require that the centerline of door jamb characters be located 60 in. from the floor. This requirement facilities reading the markings by the visually impaired.

Finally, all the standards require the marking be provided on both door jambs. This again facilitates the ability to read the markings irregardless of one's location in the car. The markings will be in the line of sight for the visually-impaired. The blind, on the other hand, can reach outside the car and tactually identify the floor at which the elevator is stopped.

Originally the blind wanted raised characters on the car station buttons. The industry pointed out that you could inadvertently register numerous calls when the tactile markings are on the buttons. Actual field-testing with the blind confirmed the industry's position and it was agreed that all markings were to be placed adjacent and to the left of the car station buttons. The 5/8 in. height was required since that is the minimum size that can best be discerned tactually by the blind and visually by persons with a visual impairment.

To assist those with visual impairments, both the door jamb and car station button markings were required to be on a contracting color background. The markings also were required to be raised or incised (recessed) a minimum of 0.030 in. This requirement allowed the blind to tactual identifying the character.

In 1985, the Georgia Institute of Technology conducted research that led to a report, "A Multidisciplinary Assessment of the State of the Art of Signage for Blind and Low Vision Persons." With one exception, the criteria established for the car stations was sustained. The exception was on incised characters. Georgia Institute found that incised characters were more difficult to identify by the blind for two reasons:

1. The engraving easily fills up with debris caused by moisture and dirt on finger tips as well as airborne dust.

2. Many people become blind due to diabetes, which causes a lessening of the tactile sensation in the fingertips.

As a result of this study, incised characters are not recognized by ANSI A117.1-1986, CABO/ANSI A117.1-1992, ICC/ANSI A117.1-1998, ADAAG and UFAS.

The studies undertaken by Dr. Edward Steinfeld in the 1970s concluded that only a small percentage of the blind could read Braille. He also concluded that those that could read Braille could read the raised characters. Thus ANSI A117.1-1980, ANSI A117.1-1986 and UFAS do not require Braille. However, the blind were able to convince the Access Board and eventually the ANSI A117 Committee that Braille was needed. Thus, ADAAG and CABO/ANSI A117.1-1992 require all tactile signage to be accompanied by Braille.

During the research for ANSI A117.1-1980, it became evident to both NEII and the ANSI A117 Committee that a standard identification of car control designations was required.

The ANSI A117 Committee recognized that this area was within the Scope of the ASME A17 Committee and requested they include same in A17.1. ASME A17.1 requires the use of the door open, door close, audible signaling device and emergency stop switch designations shown in Fig. B1.

In the early 1990s NEII became aware that despite the ASME A17.1 requirements for car control button symbols, further clarification was required. The ASME A17.1 Standards allowed the designer too much leeway. Complaints were being raised by the blind that the symbols in some cases were not clear when read tacitly. NEII formed a task group to work with the National Federation of the Blind. The findings of the task group were presented to the ASME A17 and CABO A117 Committees. Both committees have adopted those findings. The identical requirements for control button identification will appear in ASME A17.1-2000 and ICC/ANSI A117.1-1998. The requirements can be found in Fig. B1.

One word of caution regarding the emergency stop switch symbol in Fig. B 1. The complete octagon is raised 0.030 in. with the "X" on the face of the octagon not being tactile. Some manufacturers are raising the outline of the octagon and "X" in its center 0.030 in. This is not in compliance with the requirement. This only confused the tactual reader.

Fig. B1 – ICC/ANSI A117.1-1998 Table 407.2.11.2

Control Button	Tactile Symbol	Braille Message	Proportions Open circles indicate unused dots within each Braille Cell
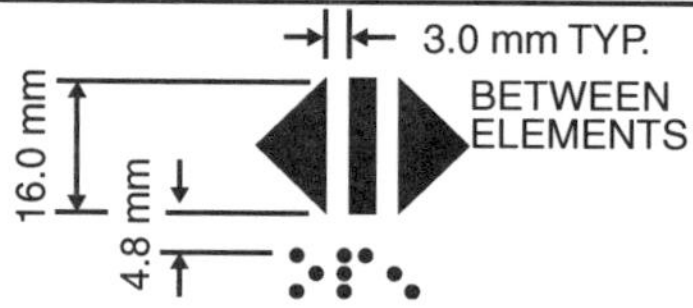 DOOR OPEN (16.0 mm, 4.8 mm, 3.0 mm TYP. BETWEEN ELEMENTS)		 OP"EN"	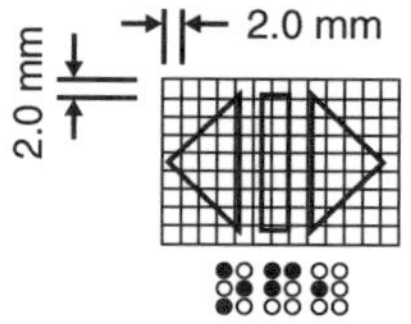 (2.0 mm)
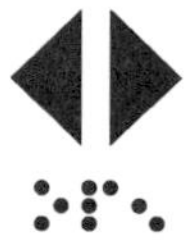 REAR/SIDE DOOR OPEN	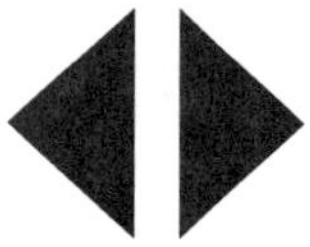	REAR/SIDE OP"EN"	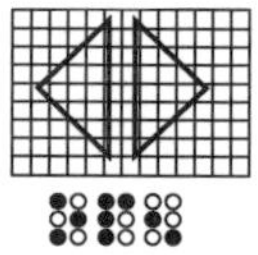
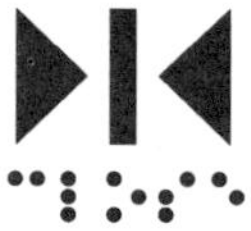 DOOR CLOSE	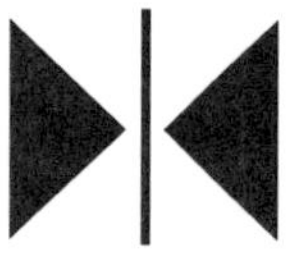	 CLOSE	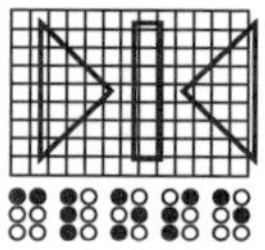
 REAR/SIDE DOOR CLOSE		 REAR/SIDE CLOSE	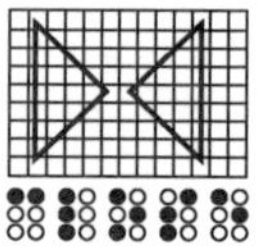
 MAIN		MA"IN"	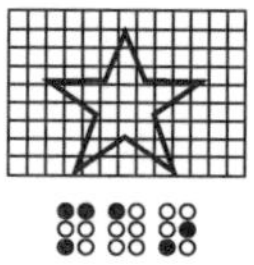
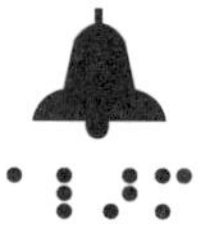 ALARM		 AL"AR"M	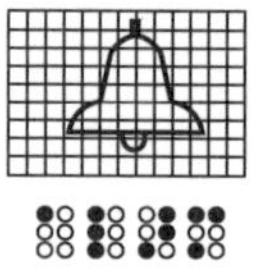
 PHONE		 PH"ONE"	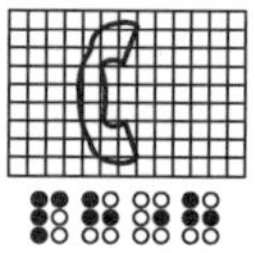
 EMERGENCY STOP (WHEN PROVIDED) X on face of octagon is not required to be tactile		 "ST"OP	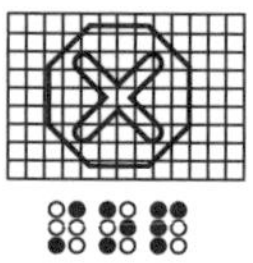

B.2 ANALYSIS OF SIGNAGE REQUIREMENTS

ADAAG, ANSI A117.1-1986, CABO/ANSI A117.1-1992, ICC/ANSI A117.1-1998 and UFAS all refer to the signage requirements in the respective standard for elevator door jamb and car control markings. The following format will be used for a Section by Section analysis of the signage regulations applicable to elevators. The **most stringent requirement** for signage is **printed in bold**. By adhering to the most stringent requirements, you will be in compliance with ADAAG, ANSI A117.1-1986, CABO/ANSI A117.1-1992, ICC/ANSI A117.1-1998 and UFAS.

State and local accessibility requirements have not been factored into the analysis. They should be adhered to, if more stringent. All of the referenced standards have Appendix material, which is considered advisory and non-binding. The Appendix material is intended to clarify the position of the organization that wrote the requirements. However, the Appendix material while not mandatory can be used to justify a good faith effort to comply with the respective requirements.

B.2.1 GENERAL

ADAAG

4.30.1* General. Signage required to be accessible by 4.1 shall comply with the applicable provisions of 4.30.

The asterisk (*) following the Section number indicates there is non-binding Appendix material.

ICC/ANSI A117.1-1998

The requirements are essentially identical for elevator signage. All sign characters are required to be presented both tactilely and visually.

CABO/ANSI A117.1-1992

The requirements are essentially identical for elevator signage. The standard does include requirement for both "accessible signage" and "tactile signage." This analysis will not address "accessible signage" as it is not applicable.

ANSI A117.1-1980

The requirements are essentially identical.

UFAS

> The requirements are essentially identical.

B.2.2 CHARACTER PROPORTION

UFAS

> 4.30.2* Character Proportion. Letters and numbers on signs shall have a width-to-height ration between 3:5 and 1:1 and a stroke width-to-height ratio between 1:5 and 1:10.

The asterisk (*) following the Section number indicates there is non-binding Appendix material.

ICC/ANSI A117.1-1998

> The requirements for characters are as follows:
> **703.2.4.1 Case. Characters shall be uppercase.**
> **703.2.4.2 Style. Characters shall be sans serif. Characters shall not be italic, oblique, script, highly decorative, or of other unusual forms.**
> **703.2.4.3 Width. Character width shall be 55 percent minimum and 110 percent maximum of the height of the character, with the width based on the uppercase letter "O" and the height based on the uppercase letter "I."**
> **703.2.4.4 Height. Character height, measured vertically from the baseline of the character, shall be • inch (16 mm) minimum, and 2 inches (51 mm) maximum, based on the uppercase letter "I". Stroke Thickness. Characters with rectangular cross sections shall have a stroke thickness, which is 10 percent minimum, and 15 percent maximum, of the height of the character, based on the uppercase letter "I". Characters with other cross sections shall have a stroke thickness at the base of the cross sections which is 10 percent minimum, and 30 percent maximum, of the height of the character, and a stroke thickness at the top of the cross sections which is 15 percent maximum of the height of the character, based on the uppercase letter "I".**

CABO/ANSI A117.1-1992

> The requirements are essentially identical except it specifies utilizing an upper-case "X" for measurement.

ANSI A117.1-1986

The requirements are identical to ANSI A117.1-1992.

UFAS

The requirements are identical to ADAAG.

B.2.3 TACTILE CHARACTERS AND SYMBOLS

ADAAG

4.30.4* Raised and Brailled Characters and Pictorial Symbol Signs (Pictograms). Letters and numerals shall be raised 1/32 in. upper case, sans serif or simple serif type **and shall be accompanied with Grade 2 Braille. Raised character shall be at least 5/8 in. (16 mm) high, but no higher than 2 in. (50 mm).**

The asterisk (*) following the Section number indicates there is non-binding Appendix material.

ICC/ANSI A117.1-1998

The requirements are essentially identical with the following additional requirements. **Braille provided on elevator car controls shall be separated $^{3}/_{16}$ inch (4.8 mm) minimum either directly below or adjacent to the corresponding raised characters or symbols.**

The Braille identification of car control buttons shall be as shown in Fig. B1, Braille dots shall have a domed or rounded shape and shall conform to the specifications in Fig. B2.

Fig. B2 – Measurement Range for Standards Sign Braille

Measurement Range For:	Minimum	Maximum
Dot base diameter	0.059 inch (1.5 mm)	0.063 inch (1.6 mm)
Distance between two dots in same cell, center to center	0.090 inch (2.3 mm)	0.100 inch (2.5 mm)
Distance between corresponding dots in adjacent cells, center to center	0.241 inch (6.1 mm)	0.300 inch (7.6 mm)
Dot height	0.025 inch (0.6 mm)	0.037 inch (0.9 mm)
Distance between corresponding dots from one cell to the cell directly below, center to center	0.395 inch (10.0 mm)	0.400 inch (10.1 mm)

CABO/ANSI A117.1-1992

The requirements are essentially identical with the following additional requirement: Braille provided in accordance with 4.10.12 shall be placed 3/16 in. (5 mm) minimum below the corresponding raised characters or symbols. Braille shall be Grade 2 and shall conform to Specification #800, National Library Service, Library of Congress.

ANSI A117.1-1986

The requirements are essentially identical, except Braille is not required.

UFAS

The requirements are essentially identical, except only sans serif characters are recognized.

B.2.4 FINISH AND CONTRAST

ADAAG

> **4.30.5* Finish and Contrast. The characters and background of signs shall be eggshell, matte, or other non glare finish. Characters and symbols shall contrast with their background - either light character on a dark background or dark characters on a light background.**

The asterisk (*) following the Section number indicates there is non-binding Appendix material.

ICC/ANSI A117.1-1998

The requirements are essentially identical except the **background must have a non-glare finish.**

CABO/ANSI A117.1-1992

The requirements are identical.

ANSI A117.1-1986

The requirements are essential identical. A background of eggshell matte or other non-glare finish is not specified.

UFAS

The requirements are identical to ANSI A117.1-1986.

OBSERVATION

The Appendix material to CABO/ANSI A117.1-1992 gives the following advise on determining finish and contrast:

> An eggshell finish (11 to 19 degree gloss on 60 degree glossimeter) is recommended. Research indicates that signs are more legible for persons with low vision when characters contrast with their background by 70 percent minimum. Contrast, in percent, is determined by:
>
> $$\text{Contrast} = [(B_1 - B_2)/B_1] \times 100$$
> where B_1 = light reflectance value (LRV) of the lighter area
> and B_2 = light reflectance value (LRV) of the darker area
>
> Note that in any application both white and black are never absolute; thus B_1 never equals 100 and B_2 is always greater than 0. The greatest readability is usually achieved through the use of light-colored characters or symbols on a dark background.

APPENDIX C
DRAFT ADA/ABA Accessibility Guidelines
Building Transportation Requirements

The following excerpts are from the ADA/ABA Accessibility Guidelines, published in the Federal Register, November 16, 1999. The reprinted material is not complete, and is only intended to assist the building transportation industry review the proposed regulations. The proposed regulations are available on the Internet at www.access-board.gov/ada-aba/guideprm.htm.

AMERICANS WITH DISABILITIES ACT SCOPING

101 Purpose
This part provides scoping and technical requirements for accessibility to sites, facilities, buildings, and elements by individuals with disabilities. These requirements are to be applied during the design, construction, and alteration of sites, facilities, buildings, and elements to the extent required by regulations issued by Federal agencies under the Americans with Disabilities Act of 1990.

102 Provisions for Adults and Children
The technical requirements in this part are based on adult dimensions and anthropometrics. This part also contains technical requirements based on children's dimensions and anthropometrics for drinking fountains, water closets, toilet compartments, lavatories and sinks, dining surfaces, and work surfaces.

103 Equivalent Facilitation
Nothing in this part is intended to prevent the use of designs or technologies as alternatives to those prescribed in this part provided they result in substantially equivalent or greater accessibility and usability.

> **Advisory 103**
> The responsibility for demonstrating equivalent facilitation in the event of a challenge rests with the covered entity. With the exception of transit facilities which are covered by regulations issued by the Department of Transportation, there is no process for certifying that an alternative design provides equivalent facilitation.

105.1 General. The standards referenced in this part and listed in 105.2 shall be considered part of the requirements of this part to the prescribed extent of each such reference. References to standards within the technical and scoping requirements shall apply to the specific edition of the reference standard listed in 105.2. Where differences occur between provisions of this part and referenced standards, the provisions of this part shall apply.

> **Advisory 105.1**
> In addition to the requirements of this document, there is an obligation to meet the requirements of any referenced standards unless there is a conflict with the guidelines. It is important to use the specific edition of the referenced standards.

105.2 Referenced Standards.
105.2.4 Safety Code for Elevators and Escalators. ASME/ANSI A17.1-1993, (including Addenda ASME/ANSI A17.1a-1994 and ASME/ANSI A17.1b-1995).

106.5 Defined Terms.
106.1 General. Terms defined in 106.5 shall have the specified meaning for purposes of this part, unless otherwise stated.
106.2 Terms Defined in Referenced Standards. Terms not defined in this section or in regulations issued by the Department of Justice and the Department of Transportation to implement the Americans with Disabilities Act but specifically defined in a referenced standard, shall have the specified meaning from the referenced standard, unless otherwise stated.
Destination-Oriented Elevator. An elevator system that provides lobby controls to select floor stops, lobby indicators designating which elevator to use and a car indicator designating the floors at which the car will stop.

203 General Exceptions
203.1 General. Sites, buildings, facilities, and elements shall be exempt from this part to the extent specified by 203.
203.6 Equipment Spaces. Spaces frequented only by service personnel for maintenance, repair, or occasional monitoring of equipment are not required to be accessible. Such spaces include but are not limited to elevator pits, elevator penthouses, mechanical, electrical, or communications equipment rooms, piping or equipment catwalks, water or sewage treatment pump rooms and stations, electric substations and transformer vaults, and highway and tunnel utility facilities.

206.2.3 Multi-Level Buildings and Facilities. Accessible routes shall connect each level, including mezzanines, in multi- level buildings and facilities.

> ### Advisory 206.2.3
> While a mezzanine may be a change in level, it is not a story for purposes of determining whether a building or facility is required to have an elevator. If an elevator is required, it must serve mezzanines. Levels in buildings and facilities without elevators must still fully comply with this document.

EXCEPTIONS: 1. An accessible route is not required to levels located above or below the accessible level in private buildings or facilities that are less than three stories or that have less than 3000 square feet per story unless the building or facility is a shopping center, a shopping mall, the professional office of a health care provider, or another type of facility as determined by the Attorney General. In addition, Exception 1 shall not apply to a terminal, depot or other station used for specified public transportation or to an airport passenger terminal.
2. An accessible route is not required to levels located above or below the accessible level in public buildings or facilities that are less than three stories and that are not open to the public if the level above or below the accessible level houses no more than five persons and is less than 500 square feet.

3. An accessible route is not required to levels located above or below the accessible level in detention and correctional facilities where accessible cells or rooms provided in accordance with 233, all common use areas serving such cells or rooms, and all public use areas are on an accessible route.

4. An accessible route is not required to levels located above or below the accessible level in residential facilities where accessible dwelling units complying with 234, all common use areas serving such dwelling units, and all public use areas are on an accessible route.

> **Advisory 206.2.3 Exception 4**
> Where common use areas are provided for the use of residents, it is presumed that all such common use areas "serve" accessible dwelling units unless use is restricted to residents occupying other dwelling units. For example, if all residents are permitted to use all laundry rooms, then all laundry rooms "serve" accessible dwelling units. However, if the laundry room on the first floor is restricted to use by residents on the first floor, and the second floor laundry room is for use by occupants of the second floor, then first floor accessible units are "served" only by laundry rooms on the first floor. In this example, an accessible route is not required to the second floor provided that all accessible units and all common use use areas serving them are on the first floor.

5. An accessible route is not required to levels located above or below the accessible level in multi-story transient lodging guest rooms provided that spaces complying with 806.2 are on an accessible route and are suitable for dual occupancy.

6. In assembly areas required to comply with 221, an accessible route is not required to serve seating where wheelchair spaces or designated aisle seats required to be on an accessible route are not provided.

7. In air traffic control towers, an accessible route is not required to serve the cab and the floor immediately below the cab.

8. In alterations to qualified historic buildings or facilities where an exception is permitted by 202.5, an accessible route from an accessible entrance to all publicly used spaces on at least the level of the accessible entrance shall be provided.

206.2.3.1 Stairs and Escalators in Existing Buildings. In alterations and additions, where an escalator or stair is provided where none existed previously and major structural modifications are necessary for such installation, an accessible route shall be provided between the levels served by the escalator or stair, unless exempted by 206.2.3.

206.6 Elevators. New passenger elevators shall comply with 407.2 or 407.3. Where multiple elevators are provided, each passenger elevator shall comply with 407.2 or 407.3.
   EXCEPTION: *Where an elevator is provided in a building or facility eligible for the exceptions to 206.2.3, the elevator shall comply with 407.2, 407.3 or 407.4.*

206.6.1 Existing Elevators. Altered elements of existing elevators shall comply with 407.5. Such elements shall also be altered in all elevators that are programmed to respond to the same hall call control as the altered elevator and shall comply with the requirements of 407.5.

206.7 Wheelchair (Platform) Lifts. Wheelchair (platform) lifts shall be permitted as a component of an accessible route in new construction as permitted by 206.7 and shall comply with 408. Wheelchair (platform) lifts provided as a component of an accessible route in an existing building or facility shall comply with 408.

206.7.1 Performance Areas and Speakers' Platforms. Wheelchair (platform) lifts shall be permitted to provide an accessible route to a performance area or a speakers' platform in an assembly occupancy.

206.7.2 Wheelchair Spaces. Wheelchair (platform) lifts shall be permitted to comply with the wheelchair space dispersion and line-of-sight requirements of 221 and 802.

206.7.3 Incidental Spaces. Wheelchair (platform) lifts shall be permitted to provide an accessible route to incidental occupiable spaces and rooms which are not open to the public and which are occupied by five persons maximum.

206.7.4 Judicial Spaces. Wheelchair (platform) lifts shall be permitted to provide an accessible route to raised judges' benches, clerks' stations, jury boxes and witness stands or to depressed areas such as the well of a court.

207 Accessible Means of Egress

207.1 General. All accessible spaces shall be provided with not less than one accessible means of egress. Where more than one means of egress is required from any accessible space, each accessible portion of the space shall be served by not less than two accessible means of egress. Accessible means of egress shall comply with 409.

> **EXCEPTION:** *Accessible means of egress are not required in alterations to existing buildings or facilities.*

207.2 Elevators. In buildings or facilities where a required accessible floor is four or more stories above or below a level of exit discharge, at least one required accessible means of egress shall be an elevator complying with 409.3.

ARCHITECTURAL BARRIERS ACT SCOOPING

F101 Purpose
This part provides scoping and technical requirements for accessibility to sites, facilities, buildings, and elements by individuals with disabilities. These requirements are to be applied during the design, construction, lease and alteration of sites, facilities, buildings, and elements to the extent required by regulations issued by Federal agencies under the Architectural Barriers Act of 1968.

F102 Provisions for Adults and Children
The technical requirements applicable to this part are based on adult dimensions and anthropometrics. In addition, technical requirements based on children's dimensions and anthropometrics for drinking fountains, water closets, toilet compartments, lavatories and sinks, dining surfaces and work surfaces are included.

F103 Modifications and Waivers
The Architectural Barriers Act authorizes the Administrator of the General Services Administration, the Secretary of the Department of Housing and Urban Development, the Secretary of the Department of Defense, and the United States Postal Service to modify or waive the accessibility standards for buildings and facilities covered by the Architectural Barriers Act on a case-by-case basis, upon application made by the head of the department, agency, or instrumentality of the United States concerned. The General

Services Administration, the Department of Housing and Urban Development, the Department of Defense, and the United States Postal Service may grant a modification or waiver only upon a determination that it is clearly necessary. Section 502(b)(1) of the Rehabilitation Act of 1973 authorizes the Access Board to ensure that modifications and waivers are based on findings of fact and are not inconsistent with the Architectural Barriers Act.

> **Advisory F103**
> The provisions for modifications and waivers differ from the guidelines issued under the Americans with Disabilities Act in that "equivalent facilitation" does not apply. There is a formal procedure for Federal agencies to request a waiver or modification of applicable standards under the Architectural Barriers Act of 1968, 42 U.S.C. section 4156.

F105 Referenced Standards

F105.1 General. The standards referenced in this part and listed in F105.2 shall be considered part of the requirements of this part to the prescribed extent of each such reference. References to standards within the technical and scoping requirements shall apply to the specific edition of the reference standard listed in F105.2. Where differences occur between provisions of this part and referenced standards, the provisions of this part shall apply.

> **Advisory F105.1**
> In addition to the requirements of this document, there is an obligation to meet the requirements of any referenced standards unless there is a conflict with the guidelines. It is important to use the specific edition of the referenced standards.

F105.2 Referenced Standards.

F105.2.4 Safety Code for Elevators and Escalators. ASME/ANSI A17.1-1993, (including Addenda ASME/ANSI A17.1a-1994 and ASME/ANSI A17.1b-1995).

F106 Definitions

F106.1 General. Terms defined in F106.5 shall have the specified meaning for purposes of this part, unless otherwise stated.

F106.2 Terms Defined in Referenced Standards. Terms not defined in this section, or in regulations issued by the Administrator of the General Services Administration, the Secretary of Defense, the Secretary of Housing and Urban Development, or the U.S. Postal Service to implement the Architectural Barriers Act but specifically defined in a referenced standard, shall have the specified meaning from the referenced standard, unless otherwise stated.

Destination-Oriented Elevator. An elevator system that provides lobby controls to select floor stops, lobby indicators designating which elevator to use and a car indicator designating the floors at which the car will stop.

F203 General Exceptions

F203.1 General. Sites, buildings, facilities, and elements shall be exempt from this part to the extent specified by F203.

F203.5 Equipment Spaces. Spaces frequented only by service personnel for maintenance, repair, or occasional monitoring of equipment are not required to be accessible. Such spaces include but are not limited to elevator pits, elevator penthouses, mechanical, electrical, or communications equipment rooms, piping or equipment catwalks, water or sewage treatment pump rooms and stations, electric substations and transformer vaults, and highway and tunnel utility facilities.

F206.2.3 Multi-Level Buildings and Facilities. Accessible routes shall connect each level, including mezzanines, in multi-level buildings and facilities.

> **Advisory F206.2.3**
> While a mezzanine may be a change in level, it is not a story for purposes of determining whether a building or facility is required to have an elevator. If an elevator is required, it must serve mezzanines. Levels in buildings and facilities without elevators must still fully comply with this document.

EXCEPTIONS: 1. An accessible route is not required to levels located above or below the accessible level in buildings or facilities that are less than three stories and that are not open to the public if the level above or below the accessible level houses no more than five persons and is less than 500 square feet.
2. An accessible route is not required to levels located above or below the accessible level in detention and correctional facilities where accessible cells or rooms provided in accordance with F233, all common use areas serving such cells or rooms, and all public use areas are on an accessible route.
3. An accessible route is not required to levels located above or below the accessible level in residential facilities where accessible dwelling units complying with F234, all common use areas serving such dwelling units, and all public use areas are on an accessible route.

> **Advisory F206.2.3 Exception 3**
> Where common use areas are provided for the use of residents, it is presumed that all such common use areas "serve" accessible dwelling units unless use is restricted to residents occupying other dwelling units. For example, if all residents are permitted to use all laundry rooms, then all laundry rooms "serve" accessible dwelling units. However, if the laundry room on the first floor is restricted to use by residents on the first floor, and the second floor laundry room is for use by occupants of the second floor, then first floor accessible units are "served" only by laundry rooms on the first floor. In this example, an accessible route is not required to the second floor provided that all accessible units and all common use areas serving them are on the first floor.

4. An accessible route is not required to levels located above or below the accessible level in multi-story transient lodging guest rooms provided that spaces complying with 806.2 are on an accessible route and are suitable for dual occupancy.
5. In assembly areas required to comply with F221, an accessible route is not required to serve seating where wheelchair spaces or designated aisle seats required to be on an accessible route are not provided.
6. In air traffic control towers, an accessible route is not required to serve the cab and the floor immediately below the cab.

7. In alterations to qualified historic buildings or facilities where an exception is permitted by F202.5, an accessible route from an accessible entrance to all publicly used spaces on at least the level of the accessible entrance shall be provided.

F206.2.3.1 Stairs and Escalators in Existing Buildings. In alterations and additions, where an escalator or stair is provided where none existed previously and major structural modifications are necessary for such installation, an accessible route shall be provided between the levels served by the escalator or stair, unless exempted by F206.2.3.

F206.6 Elevators. New passenger elevators shall comply with 407.2 or 407.3. Where multiple elevators are provided, each passenger elevator shall comply with 407.2 or 407.3.
EXCEPTION: Where an elevator is provided in a building or facility eligible for the exceptions to F206.2.3, the elevator shall comply with 407.2, 407.3 or 407.4.

F206.6.1 Existing Elevators. Altered elements of existing elevators shall comply with 407.5. Such elements shall also be altered in all elevators that are programmed to respond to the same hall call control as the altered elevator and shall comply with the requirements of 407.5.

F206.7 Wheelchair (Platform) Lifts . Wheelchair (platform) lifts shall be permitted as a component of an accessible route in new construction as permitted by F206.7 and shall comply with 408. Wheelchair (platform) lifts provided as a component of an accessible route in an existing building or facility shall comply with 408.

F206.7.1 Performance Areas and Speakers' Platforms. Wheelchair (platform) lifts shall be permitted to provide an accessible route to a performance area or a speakers' platform in an assembly occupancy.

F206.7.2 Wheelchair Spaces. Wheelchair (platform) lifts shall be permitted to comply with the wheelchair space dispersion and line-of-sight requirements of F221 and 802.

F206.7.3 Incidental Spaces. Wheelchair (platform) lifts shall be permitted to provide an accessible route to incidental occupiable spaces and rooms which are not open to the public and which are occupied by five persons maximum.

F206.7.4 Judicial Spaces. Wheelchair (platform) lifts shall be permitted to provide an accessible route to raised judges' benches, clerks' stations, jury boxes and witness stands or to depressed areas such as the well of a court.

F207 Accessible Means of Egress

F207.1 General. All accessible spaces shall be provided with not less than one accessible means of egress. Where more than one means of egress is required from any accessible space, each accessible portion of the space shall be served by not less than two accessible means of egress. Accessible means of egress shall comply with 409.
EXCEPTION: Accessible means of egress are not required in alterations to existing buildings or facilities.

F207.2 Elevators. In buildings or facilities where a required accessible floor is four or more stories above or below a level of exit discharge, at least one required accessible means of egress shall be an elevator complying with 409.3.

CHAPTER 4: ACCESSIBLE ROUTES

401 General

401.1 Scope. Accessible routes and accessible means of egress required by Chapter 2 shall comply with the applicable provisions of this chapter.

402 Accessible Routes

402.1 General. Accessible routes shall comply with 402.
402.2 Components. Accessible routes shall consist of one or more of the following components: walking surfaces with a slope not steeper than 1:20, doorways, ramps, elevators and platform (wheelchair) lifts. All components of an accessible route shall comply with the applicable portions of this chapter.

407 Elevators

407.1 General. New elevators required to be *accessible* shall comply with 407.2. New *destination-oriented elevators* required to be *accessible* shall comply with 407.3. New limited use/limited application elevators required to be *accessible* shall comply with 407.4. *Altered elements* of existing elevators shall comply with 407.5.

407.2 New Elevators. New *accessible* elevators shall comply with 407.2 and with ASME/ANSI A17.1. They shall be passenger elevators as classified by ASME/ANSI A17.1.

> **407.2.1 Automatic Operation.** Elevator operation shall be automatic. Each car shall be equipped with a self-leveling feature that will automatically bring and maintain the car at floor landings within a tolerance of 1/2 inch (13 mm) under rated loading to zero loading conditions.
>
> **407.2.2 Call Buttons.** Call buttons in elevator lobbies and halls shall be located vertically between 35 inches (890 mm) and 48 inches (1220 mm) above the floor, measured to the centerline of the button. A clear floor *space* complying with 305 shall be provided. Such call buttons shall have visible signals to indicate when each call is registered and when each call is answered. Call buttons shall be 3/4 inch (19 mm) minimum in the smallest dimension. The button that designates the up

direction shall be located above the button that designates the down direction. Buttons shall be raised or flush. Objects located beneath hall call buttons shall protrude 4 inches (100 mm) maximum into the clear floor *space*.

Figure 407.2.2

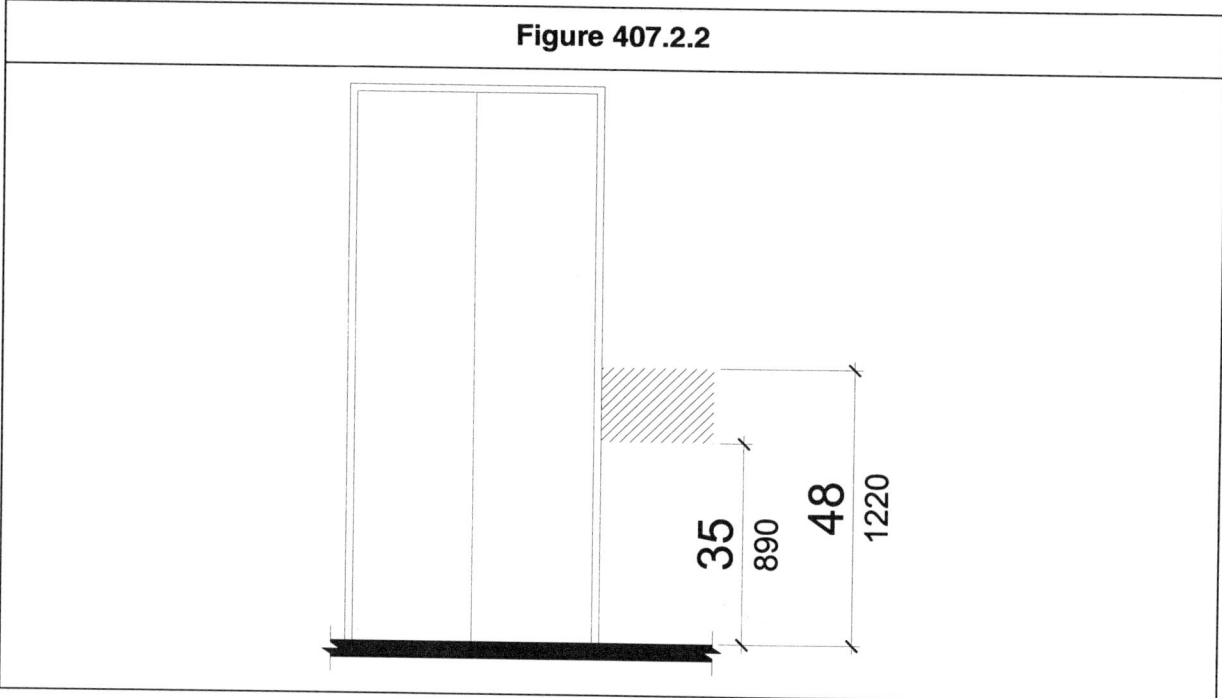

407.2.3 Hall Signals. A visible and audible signal shall be provided at each hoistway *entrance* to indicate which car is answering a call and the direction of travel. Alternatively, in-car signals shall be located in cars, visible from the floor area adjacent to the hall call buttons, and shall comply with the requirements of this section.

407.2.3.1 Audible Signals. Audible signals shall sound once for the up direction and twice for the down direction, or shall have verbal annunciators that state the word "up" or "down." Audible signals or verbal annunciators shall have a frequency of 1500 Hz maximum. The audible signal or verbal annunciator shall be 20 dBA minimum and 80 dBA maximum, measured at the hall call button.

407.2.3.2 Visible Signals. Visible signals shall comply with 407.2.3.2.

407.2.3.2.1 Height. Hall signal fixtures shall be centered at 72 inches (1830 mm) minimum above the floor or ground.

Figure 407.2.3.2.1

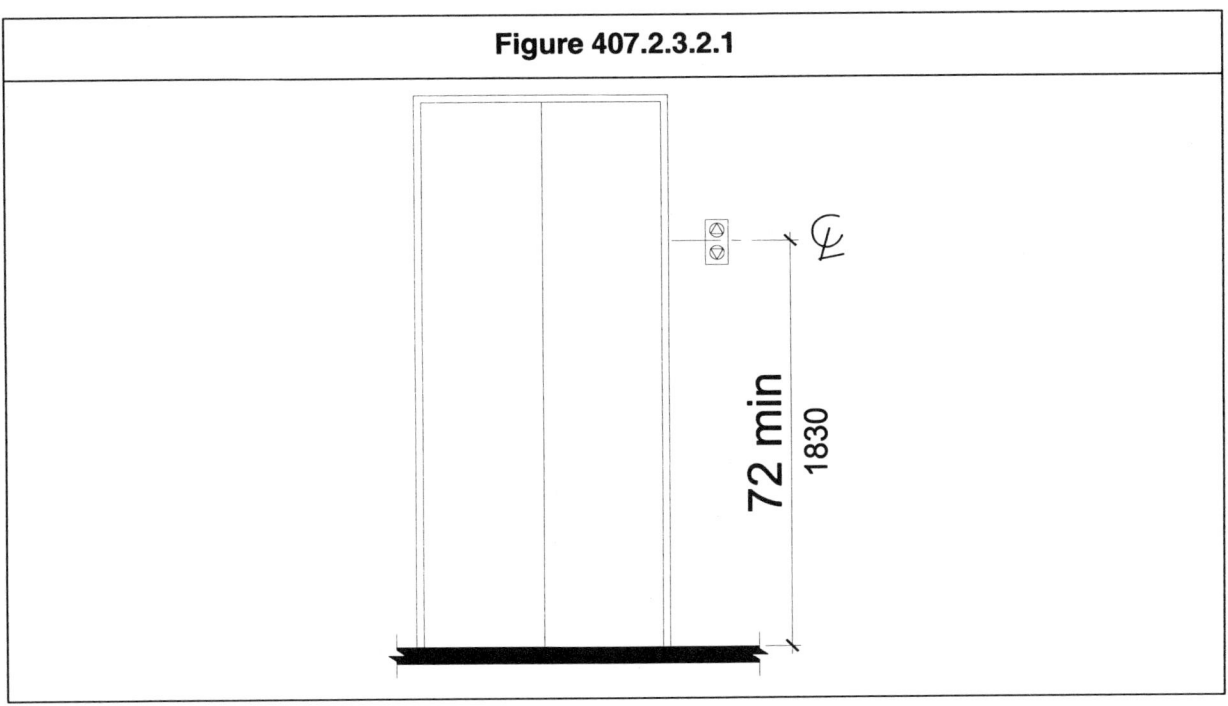

407.2.3.2.2 Size. The visible signal *elements* shall be 2-1/2 inches (64 mm) minimum measured along the vertical centerline of the *element*.

Figure 407.2.3.2.2

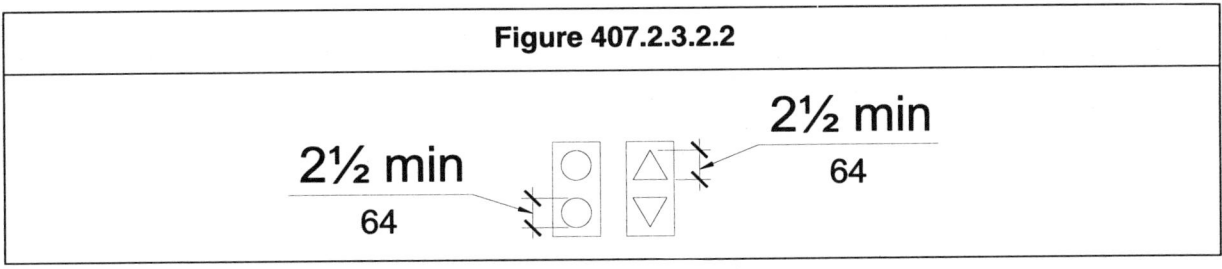

407.2.3.2.3 Visibility. Signals shall be visible from the floor area adjacent to the hall call button.

407.2.4 Tactile Signs on Hoistway Entrances. *Tactile character* and Braille floor designations shall be provided on both jambs of elevator hoistway *entrances* and shall be 60 inches (1525 mm) above the floor, measured from the baseline of the *characters*. A *tactile* star shall also be provided on both jambs at the main entry level. Such *characters* shall be 2 inches (51 mm) high and shall comply with 703.2.

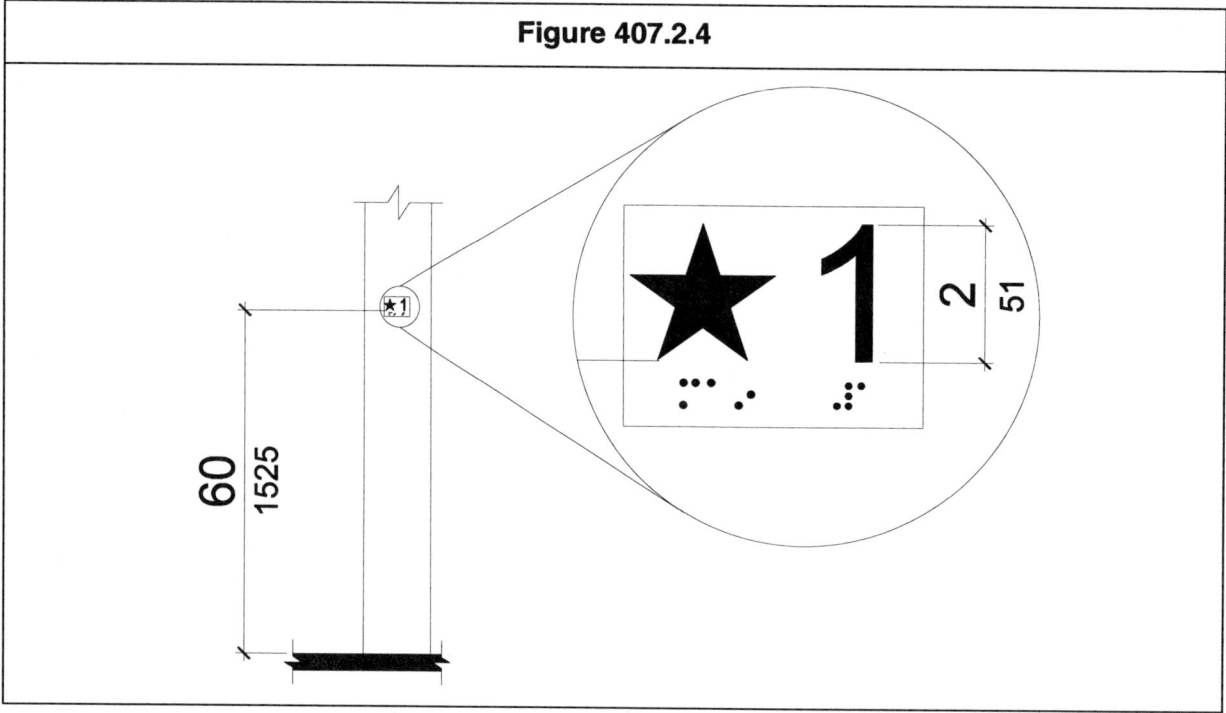

Figure 407.2.4

407.2.5 Door Operation. Elevator doors shall be the horizontal type. Elevator hoistway and car doors shall open and close automatically. Elevator doors shall be provided with a reopening device that shall stop and reopen a car door and hoistway door automatically if the door becomes obstructed by an object or person. The device shall be activated by sensing an obstruction passing through the opening at 5 inches (125 mm) and 29 inches (735 mm) above the floor. The device shall not require physical contact to be activated, although contact may occur before the door reverses. Door reopening devices shall remain effective for 20 seconds minimum.

Figure 407.2.5

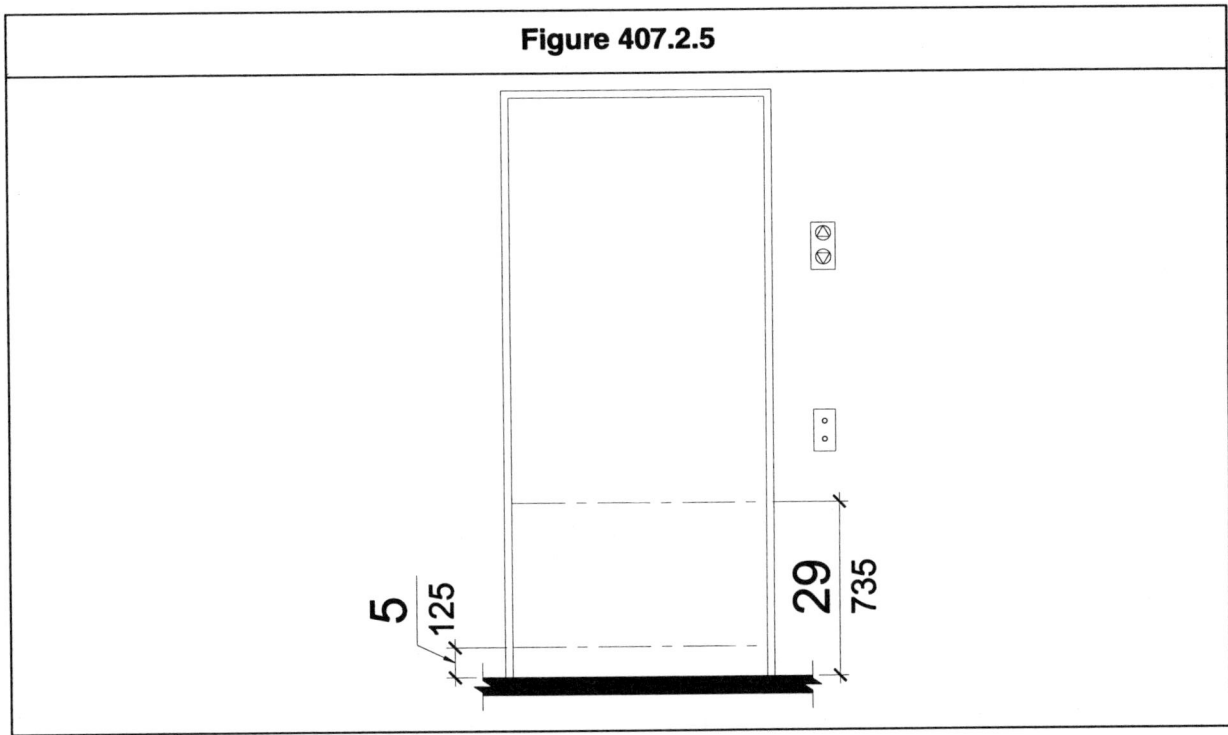

407.2.6 Door and Signal Timing for Hall Calls. The minimum acceptable time from notification that a car is answering a call or designation of which car is assigned to a lobby destination floor entry until the doors of that car start to close shall be calculated from the following equation:

$T = D/(1.5 \text{ ft/s})$ or
$T = D/(455 \text{ mm/s}) = 5$ seconds minimum

where T equals the total time in seconds and D equals the distance (in feet or millimeters) from the point in the lobby or corridor 60 inches (1525 mm) directly in front of the farthest call button controlling that car to the centerline of its hoistway door. For cars with in-car lanterns, T begins when the signal is visible from the point 60 inches (1525 mm) directly in front of the farthest hall call button and the audible signal is sounded.

407.2.7 Door Delay for Car Calls. Elevator doors shall remain fully open in response to a car call for 3 seconds minimum.

407.2.8 Inside Dimensions of Elevator Cars. Clear width of elevator doors and inside dimensions of elevator cars shall comply with Table 407.2.8.

Table 407.2.8 Elevator Door and Car Sizes

Door Location	Minimum Dimensions			
	Door Clear Width	Inside Car, Side to Side	Inside Car, Back Wall to Front Return	Inside Car, Back Wall to Inside Face of Door
Centered	42 inches (1065 mm)	80 inches (2030 mm)	51 inches (1295 mm)	54 inches (1370 mm)
Side (off-centered)	36 inches (915 mm)[1]	68 inches (1725 mm)	51 inches (1295 mm)	54 inches (1370 mm)
Any	36 inches (915 mm)[1]	54 inches (1370 mm)	80 inches (2030 mm)	80 inches (2030 mm)
Any	36 inches (915 mm)[1]	60 inches (1525 mm)[2]	60 inches (1525 mm)[2]	60 inches (1525 mm)[2]

1. A tolerance of minus 5/8 inch (16 mm) is permitted.
2. Other car configurations that provide a *wheelchair* turning *space* complying with 304 with the door closed are permited.

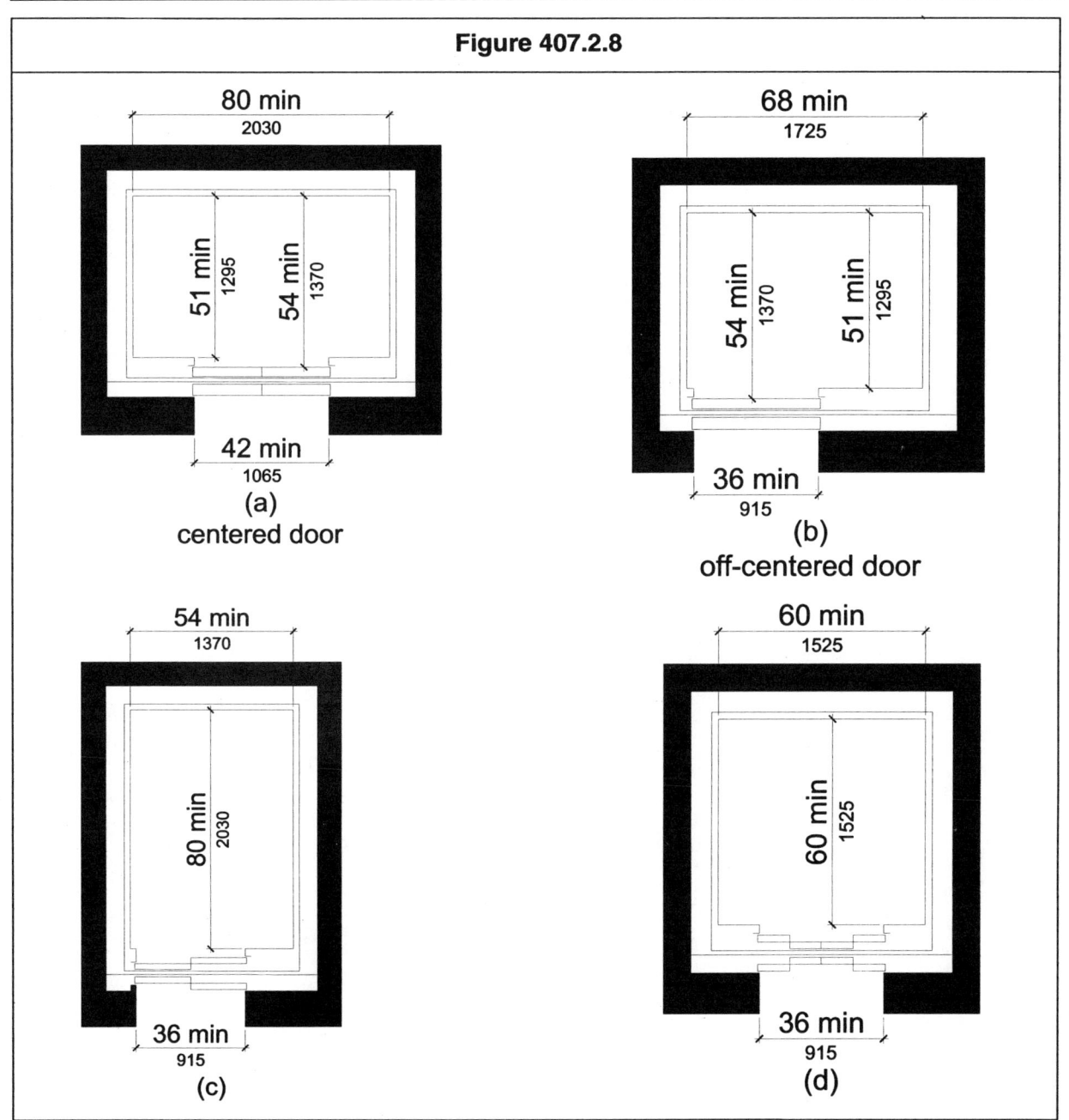

Figure 407.2.8

407.2.9 Floor Surfaces. Floor surfaces in elevator cars shall comply with 302. The clearance between the car platform sill and the edge of any hoistway landing shall be 1-1/4 inch (32 mm) maximum.

407.2.10 Illumination Levels. The level of illumination at the car controls, platform, car threshold and car landing sill shall be 5 footcandles (54 lux) minimum.

407.2.11 Car Controls. Elevator controls shall comply with 407.2.11.

407.2.11.1 Buttons. Buttons shall be 3/4 inch (19 mm) minimum in their smallest dimension. Buttons shall be raised or flush. Buttons shall be arranged with numbers in ascending order. When two or more columns of buttons are provided they shall read from left to right. Keypads, where provided, shall be in a standard telephone keypad arrangement.

Figure 407.2.11.1

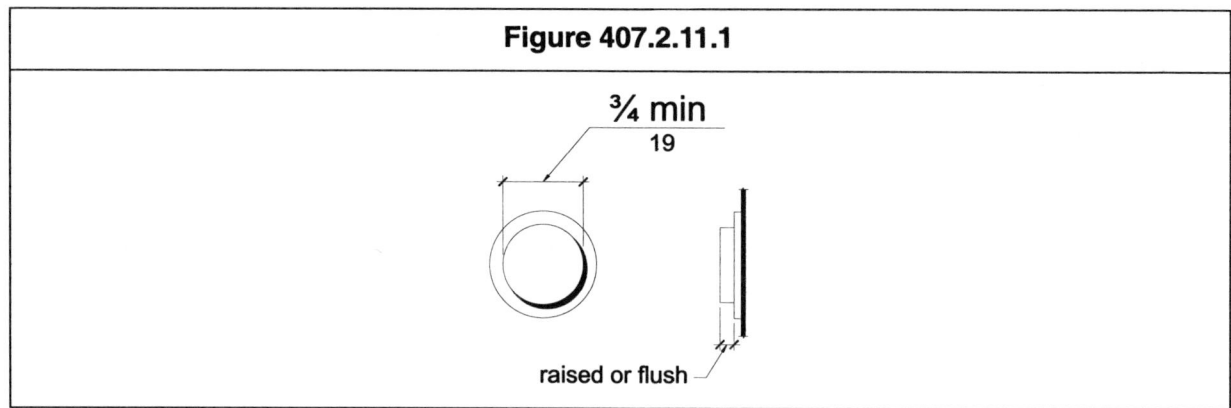

407.2.11.2 Designations and Indicators for Control Buttons. Control buttons shall be identified by *tactile characters* complying with 703.2. *Characters* and Braille shall be placed immediately to the left of the button to which the designations apply. The control button for the main entry floor and control buttons, other than remaining buttons with floor designations, shall be identified with *tactile* symbols as shown in Table 407.2.11.2. Buttons with floor designations shall be provided with visible indicators to show that a call has been registered. The visible indication shall extinguish when the car arrives at the designated floor. Where provided, telephone-style keypad buttons shall be identified by *tactile characters* complying with 703.2 except that Braille is not required. *Characters* shall be centered on the corresponding keypad button.

Table 407.2.11.2 Elevator Control Button Identification

Control Button	Tactile Symbol	Braille Message
Emergency Stop	⊗	"ST"OP" Three cells
Alarm	🔔	AL"AR"M Four cells

Table 407.2.11.2 Elevator Control Button Identification

Control Button	Tactile Symbol	Braille Message
Door Open	◀||▶	OP"EN" Three cells
Door Close	▶|◀	CLOSE Five cells
Main Entry Floor	★	MA"IN" Three cells
Phone	☎	PH"ONE" Four cells

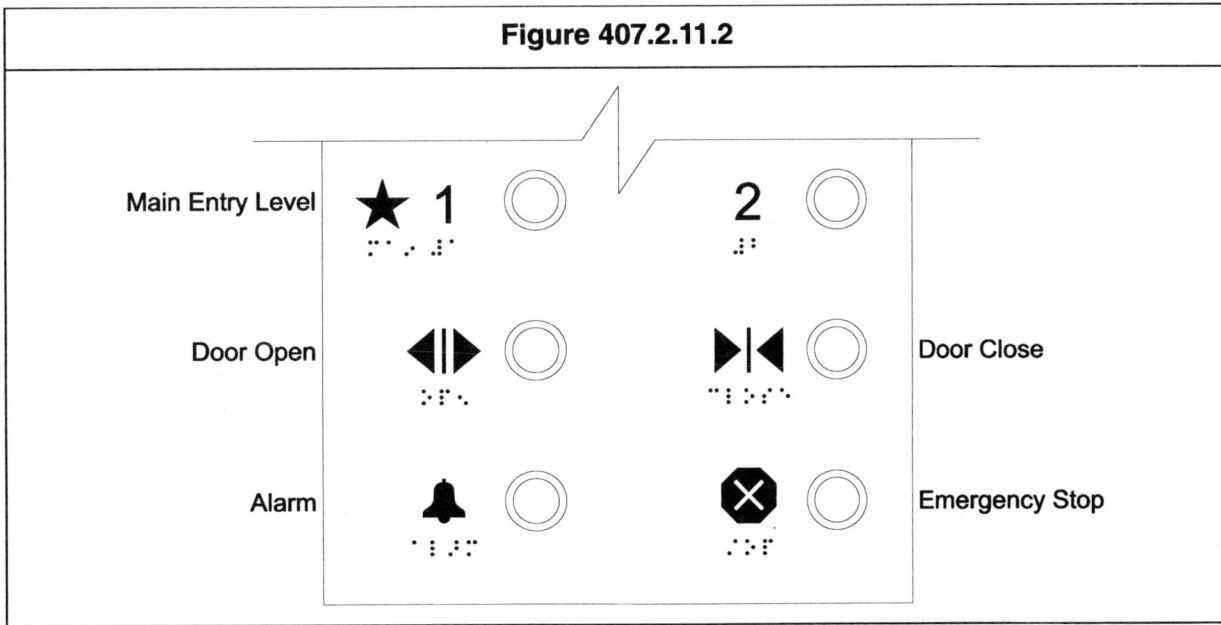

Figure 407.2.11.2

407.2.11.3 Height. Buttons with floor designations shall be located within one of the reach ranges specified in 308. Emergency controls, including the emergency alarm, shall be grouped at the bottom of the panel. Emergency control buttons shall have their centerlines 35 inches (890 mm) minimum above the floor.

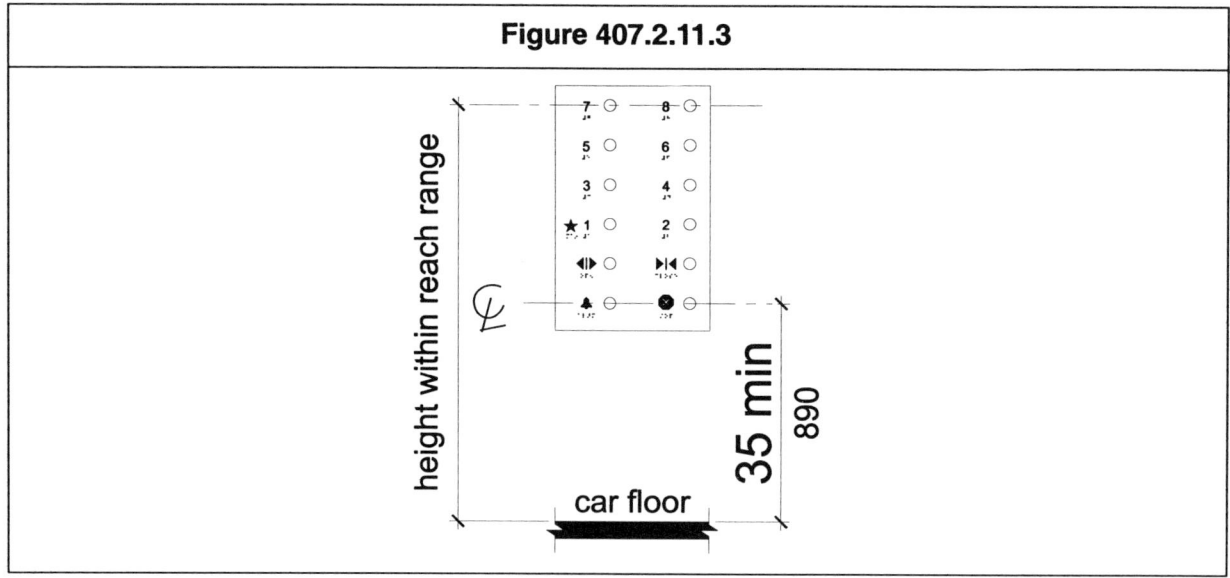

Figure 407.2.11.3

407.2.11.4 Location. Controls shall be located to accommodate a forward reach or side reach as specified in 308.

407.2.12 Car Position Indicators. In elevator cars, audible and visible car location indicators shall be provided.

407.2.12.1 Visible Indicators. Indicators shall be located above the car control panel or above the door. Numerals shall be 1/2 inch (13 mm) high minimum. As the car passes a floor and when a car stops at a floor served by the elevator, the corresponding *character* shall illuminate.

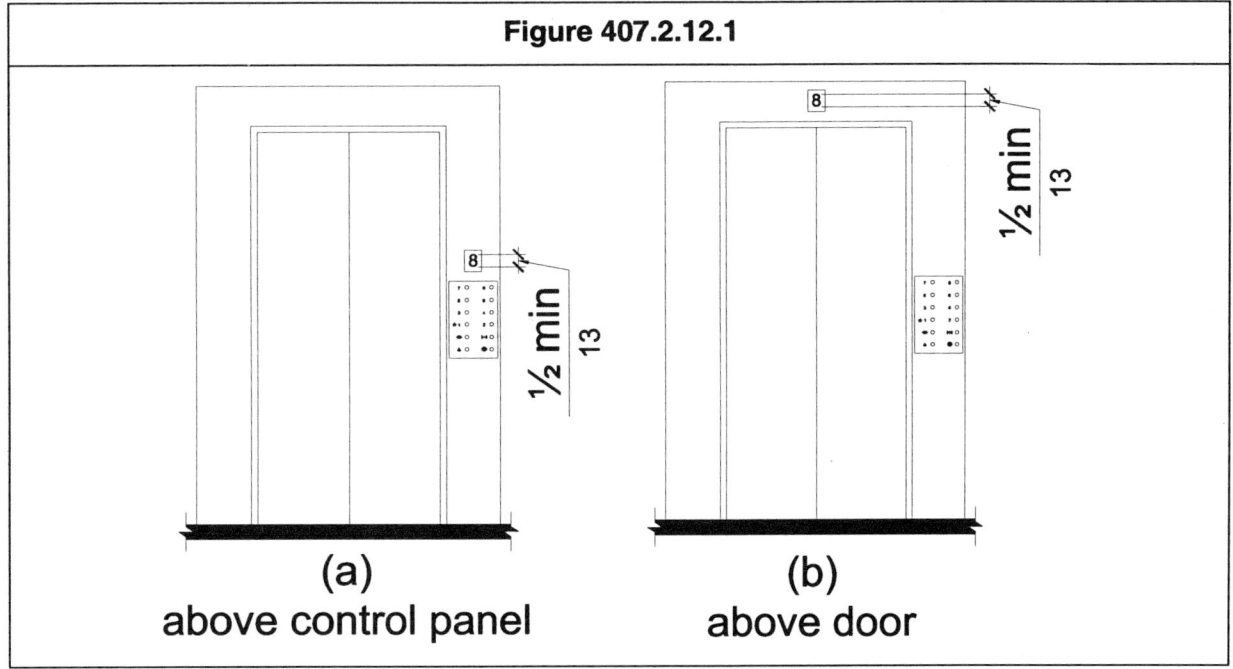

Figure 407.2.12.1

(a) above control panel

(b) above door

407.2.12.2 Audible Indicators. The audible signal shall be 20 dBA minimum and 80 dBA maximum, measured at the annunciator, and shall have a frequency of 1500 Hz maximum. The signal shall be an automatic verbal announcement which announces the floor at which the car has stopped.

> **EXCEPTION:** For elevators that have a rated speed of 200 feet per minute (1 m/s) or less, an audible signal with a frequency of 1500 Hz maximum which sounds as the car passes or stops at a floor served by the elevator shall be permitted.

407.2.13 Emergency Communications. Emergency two-way communication systems between the elevator car and a point outside the hoistway shall comply with ASME/ANSI A17.1. The highest *operable part* of a two-way communication system shall be 48 inches (1220 mm) maximum above the floor. The device shall be identified by *tactile characters* complying with 703.2 located adjacent to the device. If the system uses a handset, the cord from the panel to the handset shall be 29 inches (735 mm) long minimum. The car emergency signaling device shall not be limited to voice communication. If instructions for use are provided, essential information shall be presented in both *tactile* and visual form.

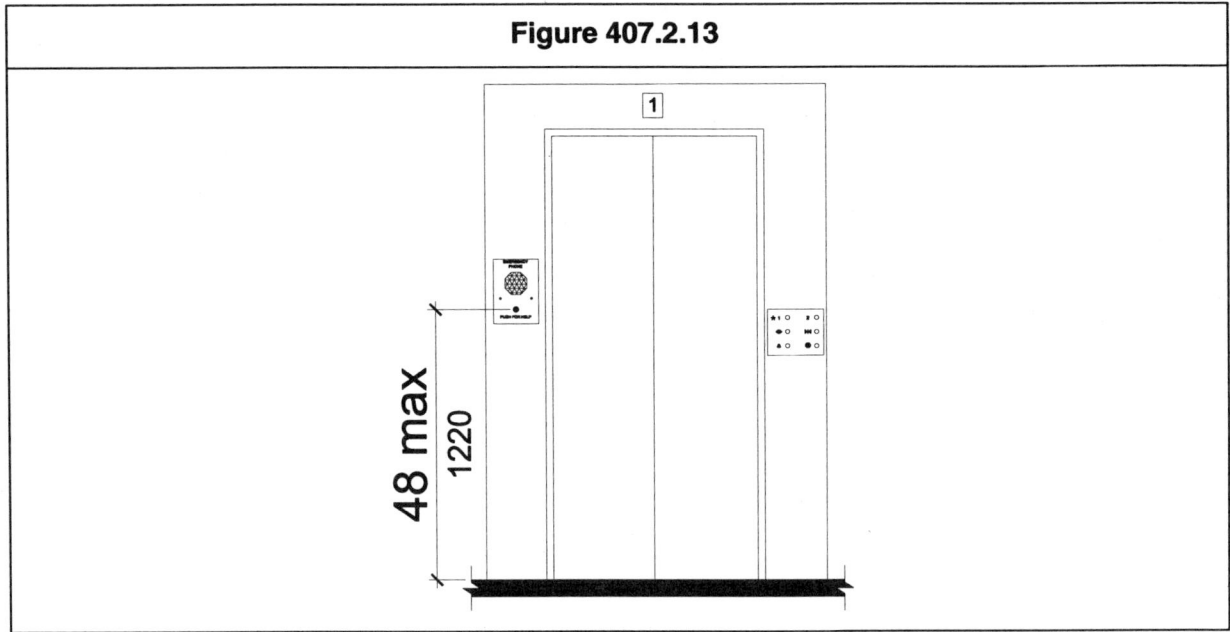

Figure 407.2.13

407.3 New Destination-Oriented Elevators. New *accessible destination-oriented elevators* shall comply with 407.2.1, 407.2.4 through 407.2.10, and 407.2.13. Such elevators shall also comply with 407.3 and ASME/ANSI A17.1. They shall be passenger elevators as classified by ASME/ANSI A17.1.

407.3.1 Call Buttons. Call buttons shall be located vertically between 35 inches (890 mm) and 48 inches (1220 mm) above the floor, measured to the centerline of the button. A clear floor *space* complying with 305 shall be provided. Call buttons shall be 3/4 inch (19 mm) minimum in the smallest dimension. Buttons shall be raised or flush. Objects located beneath hall call buttons shall protrude 4 inches (100 mm) maximum into the clear floor *space*. A keypad or other means for the entry of destination information shall be provided. Keypads, where provided, shall be in a standard telephone keypad arrangement. Visible and audible signals which indicate which elevator car to enter shall be provided.

407.3.2 Hall Signals. A visible and audible signal shall be provided to indicate a car destination corresponding with 407.3.1. The audible tone and verbal announcement shall be the same as those given at the call button or call button keypad. Each elevator in a bank shall have audible and visible means for differentiation.

407.3.2.1 Visible Signals. Visible signals shall comply with 407.3.2.1.

407.3.2.1.1 Height. Hall signal fixtures shall be centered at 72 inches (1830 mm) minimum above the floor or ground.

407.3.2.1.2 Size. The visible signal *elements* shall be 2-1/2 inches (64 mm) minimum measured along the vertical centerline of the *element*.

407.3.2.1.3 Visibility. Signals shall be visible from the floor area adjacent to the hoistway *entrance*.

407.3.3 Car Controls. Emergency controls, including the emergency alarm, shall have their centerlines 35 inches (890 mm) minimum and 48 inches (1220 mm) maximum above the floor. Buttons shall be 3/4 inch (19 mm) minimum in their smallest dimension. Buttons shall be raised or flush. Controls shall be located to accommodate a forward reach or side reach as specified in 308.

407.3.4 Car Position Indicators. In elevator cars, audible and visible car location indicators shall be provided.

407.3.4.1 Visible Indicators. Indicators shall be located above the car control panel or above the door. Numerals shall be 1/2 inch (13 mm) high minimum. A display shall be provided in the car with visible indicators to show car destinations. The visible indicators shall extinguish when the call has been answered.

407.3.4.2 Audible Indicators. An automatic verbal announcement which announces the floor at which the car has stopped shall be provided. The announcement shall be 20 dBA minimum and 80 dBA maximum, measured at the annunciator.

407.3.5 Elevator Car Identification. In addition to the *tactile signs* required by 407.2.4, a *tactile* elevator car identification shall be placed immediately below the hoistway *entrance* floor designation. The *characters* shall be 2 inches (51 mm) high and shall comply with 703.2.

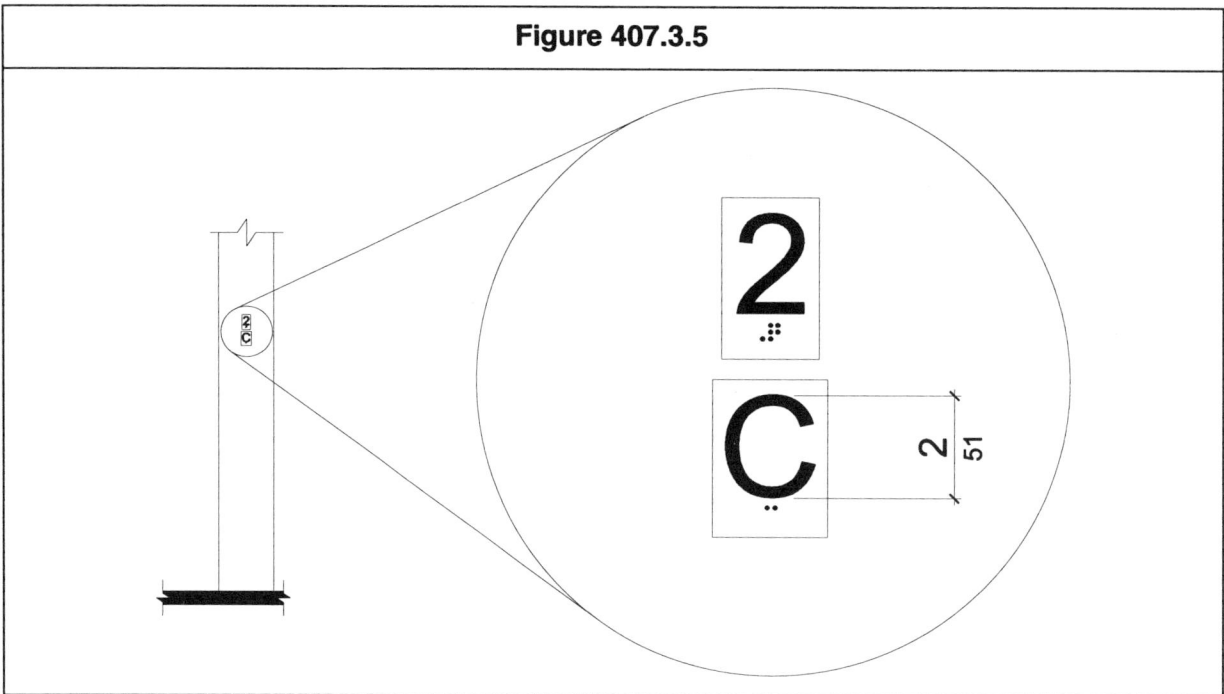

Figure 407.3.5

407.4 New Limited-Use/Limited-Application Elevators. New *accessible* limited-use/limited application elevators shall comply with 407.4 and shall comply with ASME/ANSI A17.1, Part XXV.

407.4.1 Automatic Operations. Elevator operation shall be automatic. Each car shall automatically stop at a floor landing within a tolerance of 1/2 inch (13 mm) under rated loading to zero loading conditions.

407.4.2 Call Buttons. Call buttons in elevator lobbies and halls shall be located vertically between 35 inches (890 mm) and 48 inches (1220 mm) above the floor, measured to the centerline of the button. Such call buttons shall have visible signals to indicate when each call is registered and when each call is answered. Call buttons shall be 3/4 inch (19 mm) minimum in the smallest dimension, and shall be raised or flush. The button that designates the up direction shall be located above the button that designates the down direction. Objects located beneath hall call buttons shall protrude into the floor area adjacent to the hoistway *entrance* 4 inches (100 mm) maximum.

407.4.3 Hall Signals. A visible and audible signal complying with 407.2.3 shall be provided in the car or at each hoistway *entrance* to indicate the direction of travel.

407.4.4 Tactile Signs on Hoistway Entrances. *Tactile character* and Braille floor designations shall be provided on both jambs of elevator hoistway *entrances* and shall be 60 inches (1525 mm) above the floor measured from the baseline of the *characters*. Such *characters* shall be 2 inches (51 mm) high minimum and shall comply with 703.2.

407.4.5 Door Operation. Elevator hoistway doors shall be either swinging or horizontally sliding type. Elevator hoistway and car doors shall open and close automatically. Horizontally sliding type hoistway and car doors shall comply with 407.2.5. Swinging hoistway doors shall conform to 404. Swinging doors shall be low-energy power-operated and shall comply with ANSI/BHMA A156.1.9. Power-operated swing doors shall remain open for 20 seconds minimum when activated.

407.4.6 Inside Dimensions of Elevator Cars. Elevator cars shall provide a clear width of 42 inches (1065 mm) minimum and a clear depth of 54 inches (1370 mm) minimum. For installations in existing *buildings* or *facilities*, elevator cars shall provide a clear width of 36 inches (915 mm) minimum, a clear depth of 54 inches (1370 mm) minimum, and a net clear platform area of 15 square feet (1.4 m^2) minimum. Car doors shall be positioned at the narrow end of the car and shall provide a clear width of 32 inches (815 mm) minimum.

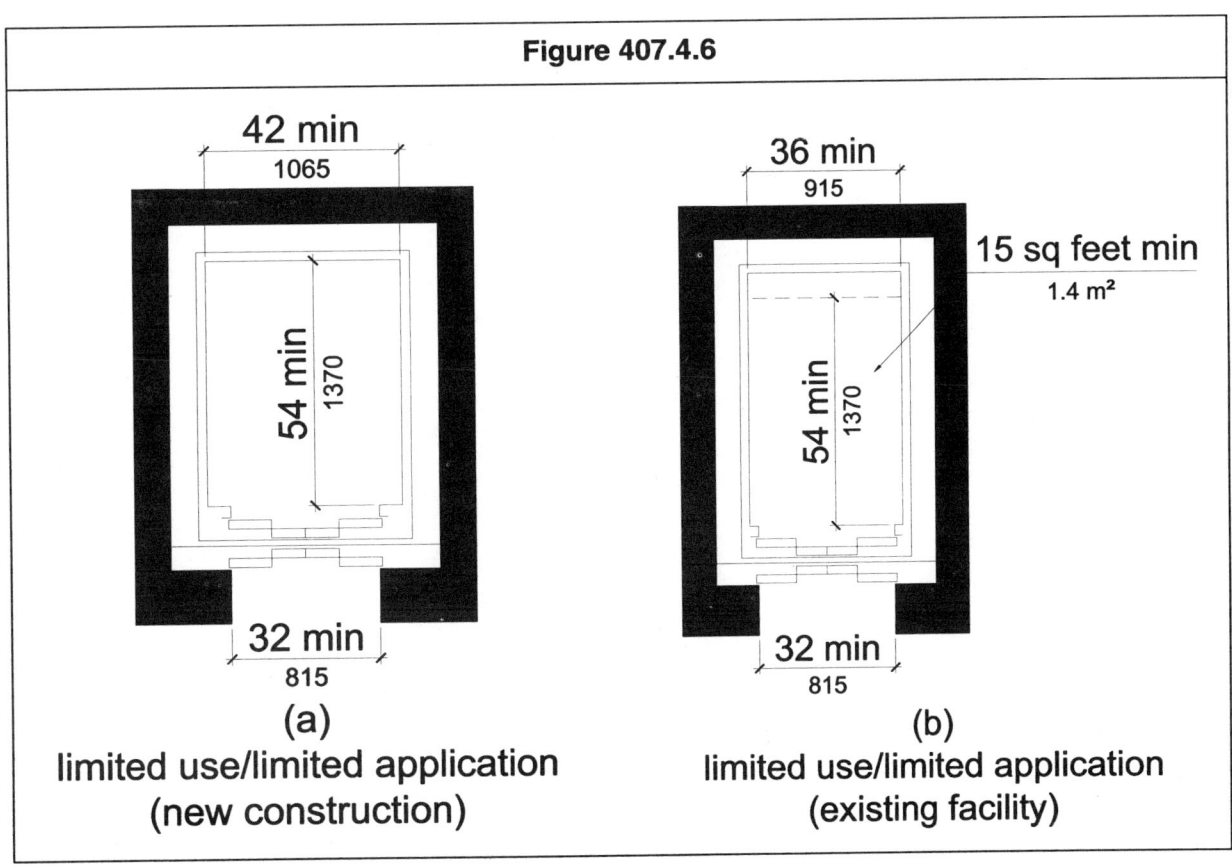

Figure 407.4.6

(a) limited use/limited application (new construction)

(b) limited use/limited application (existing facility)

407.4.7 Floor Surfaces. Floor surfaces in elevator cars shall comply with 302. The horizontal distance between the car platform sill and the edge of any hoistway landing shall be 1-1/4 inches (32 mm) maximum.

407.4.8 Illumination Levels. The level of illumination at the car controls, platform, and car threshold and landing sill shall be 5 footcandles (53.8 lux) minimum.

407.4.9 Car Controls. Elevator controls shall comply with 407.2.11. Controls shall be centered on a side wall and shall comply with 309.

407.4.10 Emergency Communications. Car emergency signaling devices complying with 407.2.13 shall be provided.

407.5 Existing Elevators. *Altered elements* of existing *destination-oriented elevators* shall comply with 407.3. *Altered elements* of existing limited-use/limited-application elevators shall comply with 407.4. *Altered elements* of all other existing elevators shall comply with 407.2.1, 407.2.4, 407.2.6, 407.2.7, 407.2.9, 407.2.10 and 407.2.13 and with 407.5 or shall comply with 407.2. They shall be passenger elevators as classified by ASME/ANSI A17.1.

407.5.1 Call Buttons. Call buttons in elevator lobbies and halls shall be located vertically between 35 inches (890 mm) and 54 inches (1370 mm) above the floor, measured to the centerline of the button. A clear floor or ground *space* complying with 305 shall be provided. The button that designates the up direction shall be located above the button that designates the down direction. Keypad controls, if provided, shall comply with 407.2.11.

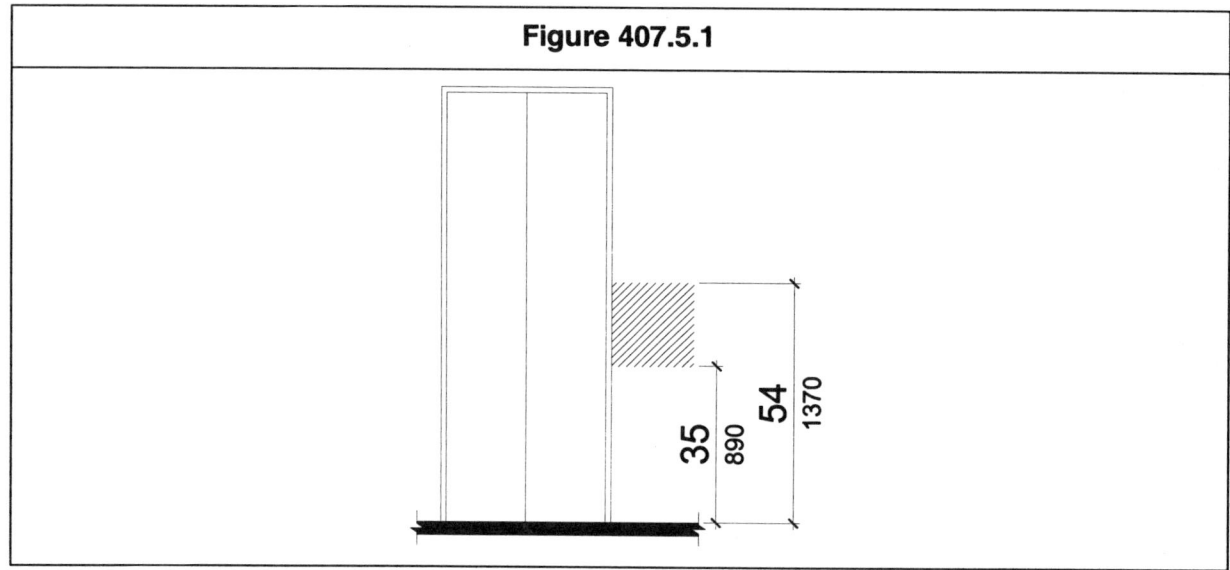

Figure 407.5.1

407.5.2 Hall Signals. A visible and audible signal at each hoistway *entrance* to indicate which car is answering a call or in-car signals complying with 407.2.3 shall be provided. Audible signals shall sound once for the up direction and twice for the down direction, or shall have verbal annunciators that state the word "up" or "down". If new hall signals are installed, they shall comply with 407.2.3.

407.5.3 Door Operation. Power-operated horizontally sliding car and hoistway doors opened and closed by automatic means shall comply with 407.2.5. Existing manually operated hoistway swing doors shall comply with 404.2.3 and 404.2.9. A power-operated car door that opens and maintains a 32 inches (815 mm) minimum clear width shall be provided. Closing of the car door shall not be initiated until the hoistway door is closed. Car gates are prohibited.

407.5.4 Inside Dimensions of Elevator Cars. The inside dimensions of elevator cars shall comply with 407.2.8.

> **EXCEPTION:** This requirement shall not apply to existing elevator car configurations that provide a clear floor area of 16 square feet (1.5 m^2) minimum, and provide 54 inches (1370 mm) minimum inside clear depth and 36 inches (915 mm) minimum clear width.

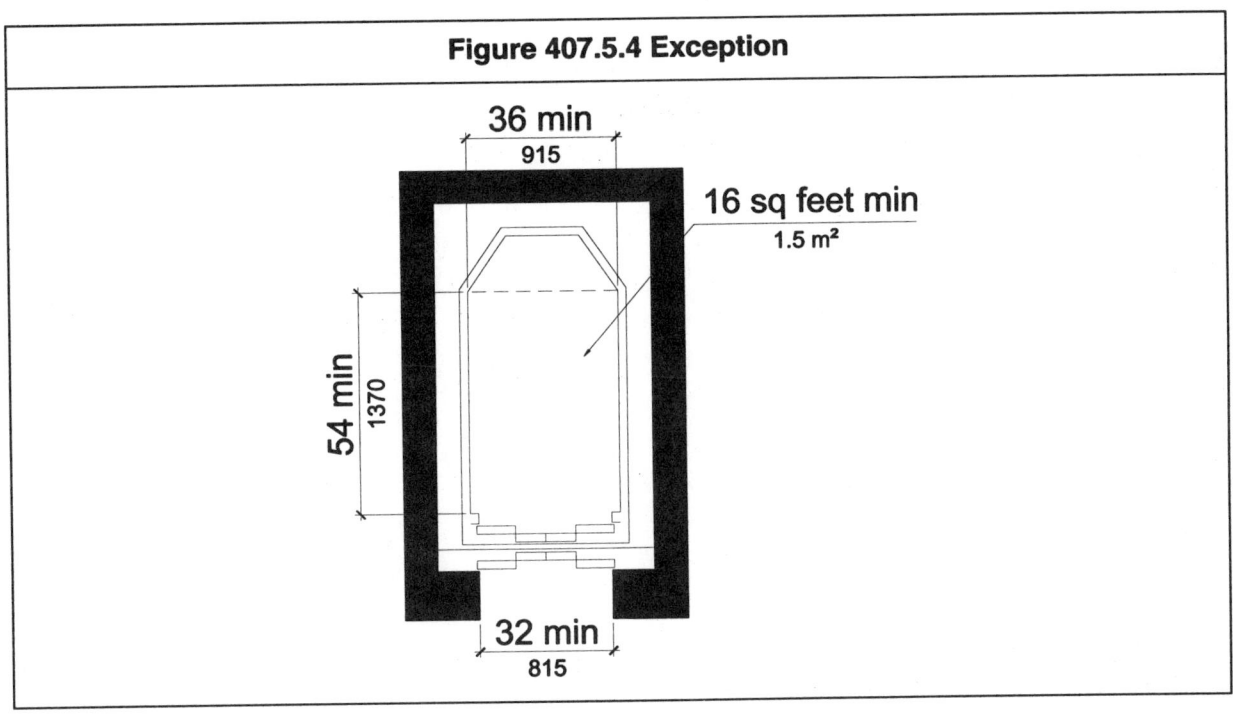

Figure 407.5.4 Exception

407.5.5 Car Controls. Elevator controls shall comply with 407.5.5.

407.5.5.1 Buttons. Control buttons shall be 3/4 inch (19 mm) minimum in their smallest dimension. Control buttons shall be raised, flush or recessed. Where the car operating panel is changed, control buttons shall comply with 407.2.11.1.

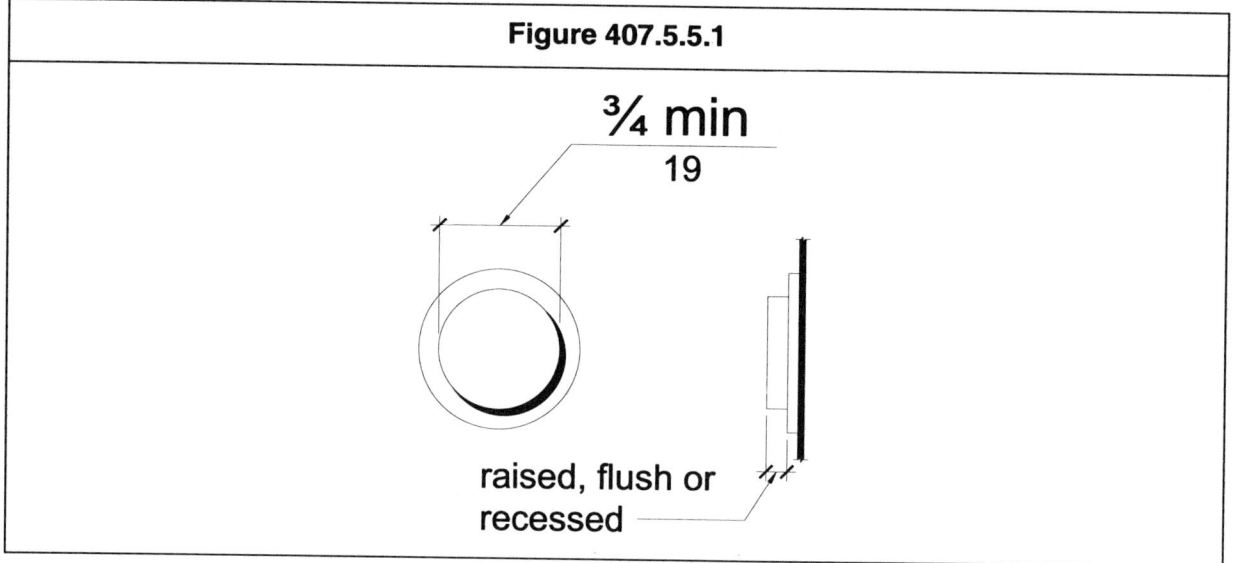

Figure 407.5.5.1

407.5.5.2 Designations and Indicators for Control Buttons. Control buttons shall comply with 407.2.11.2.

> **EXCEPTION:** Where *space* on an existing car operating panel precludes *tactile* markings to the left of the controls, markings shall be placed as near to the control as possible.

407.5.5.3 Height. Floor buttons shall be located 54 inches (1370 mm) maximum above the floor for parallel approach and 48 inches (1220 mm) maximum for front approach. Where the panel is changed, it shall comply with 407.2.11.3.

407.5.5.4 Operating Panels. Where a new car operating panel complying with 407.2.11 is provided, existing car operating panels shall not be required to comply with 407.2.11.

407.5.6 Car Position Indicators. Where a new car position indicator is provided, the indicator shall comply with 407.2.12.

407.5.7 Identification. *Accessible* elevators shall be clearly identified with the International Symbol of Accessibility complying with 703.7, unless all elevators in the *building* or *facility* are *accessible*.

408 Wheelchair (Platform) Lifts

408.1 General. *Wheelchair* (platform) lifts shall comply with ASME/ANSI A17.1 and with 302, 305 and 309. *Wheelchair* (platform) lifts shall not be attendant-operated and shall provide unassisted entry and exit from the lift.

> **Advisory 408.1**
>
> Inclined stairway chairlifts and inclined and vertical platform lifts are available for short-distance vertical transportation. Because an *accessible route* requires an 80 inch (2030 mm) vertical clearance, care should be taken in selecting lifts as they may not be equally suitable for use by *wheelchair* users and standees. If a lift does not provide 80 inch (2030 mm) vertical clearance, it cannot be considered part of an *accessible route* in new construction.

408.2 Doors and Gates. Lifts shall have low-energy power-operated doors or gates complying with 404.3. Doors and gates shall remain open for 20 seconds minimum. End doors shall be 32 inches (815 mm) minimum clear width. Side doors shall be 42 inches (1065 mm) minimum clear width.

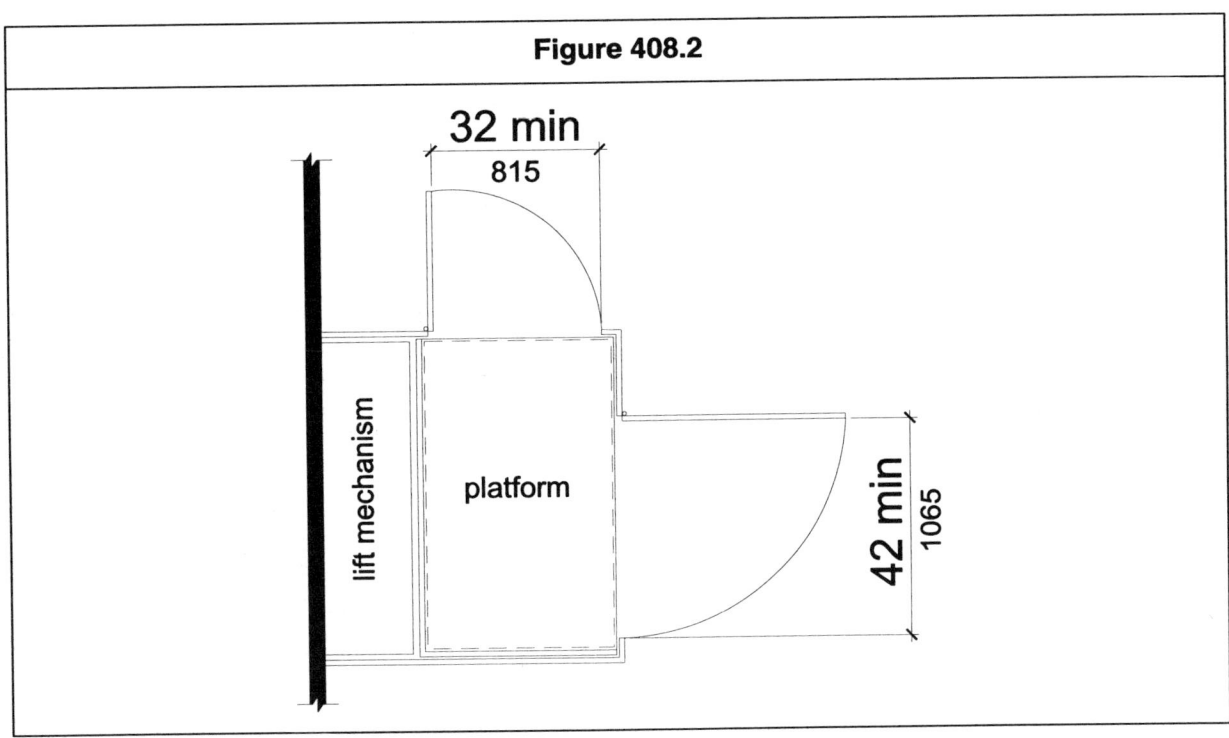

Figure 408.2

EXCEPTION: Lifts having doors or gates on opposite sides shall be permitted to have self-closing manual doors or gates.